CHAPTER 2

WIRING TECHNIQUES

LEARNING OBJECTIVES

Upon completing this chapter, you should be able to:

1. State the basic requirements for any splice and terminal connection, including the preferred wire-stripping method.

2. State the reason the ends of the wire are clamped down after a Western Union splice has been made.

3. Explain the major advantage of the crimped terminal over the soldered terminal.

4. Name the two types of insulation commonly used for noninsulated splices and terminal lugs.

5. State an advantage of using preinsulated terminal lugs and the color code used for each.

6. Explain the procedures for crimping terminal lugs with a hand crimp tool.

7. Recall the physical description and operating procedures for the HT-900B/920B compressed air/nitrogen heating tool.

8. Recall the safety precautions for using the compressed air/nitrogen heating tool.

9. Recall the procedures, precautions, and tools associated with soldering.

10. Explain the procedures and precautions for tinning wire.

11. Recall the types of soldering irons and their uses.

12. State the purposes and required properties of flux.

13. State the purpose for lacing conductors.

14. Recall when double lacing of wire bundles is required.

15. Recall the requirements for using spot ties.

WIRING TECHNIQUES

This chapter will assist you in learning the basic skills of proper wiring techniques. It explains the different ways to terminate and splice electrical conductors. It also discusses various soldering techniques that will assist you in mastering the basic soldering skills. The chapter ends with a discussion of the procedure to be followed when you lace wire bundles within electrical and electronic equipment.

CONDUCTOR SPLICES AND TERMINAL CONNECTIONS

Conductor splices and connections are an essential part of any electrical circuit. When conductors join each other or connect to a load, splices or terminals must be used. Therefore, it is important that they be properly made. Any electrical circuit is only as good as its weakest link. The basic requirement of any splice or connection is that it be both mechanically and electrically as sound as the conductor or device with which it is used. Quality workmanship and materials must be used to ensure lasting electrical contact, physical strength, and insulation. The most common methods of making splices and connections in electrical cables is explained in the discussion that follows.

INSULATION REMOVAL

The preferred method of removing insulation is with a wire-stripping tool, if available. A sharp knife may also be used. Other typical wire strippers in use in the Navy are illustrated in figure 2-1. The hot-blade, rotary, and bench wire strippers (views A, B, and C, respectively) are usually found in shops where large wire bundles are made. When using any of these automatic wire strippers, follow the manufacturer's instructions for adjusting the machine; this avoids nicking, cutting, or otherwise damaging the conductors. The hand wire strippers are common hand tools found throughout the Navy. The hand wire strippers (view D of figure 2-1) are the ones you will most likely be using. Wire strippers vary in size according to wire size and can be ordered for any size needed.

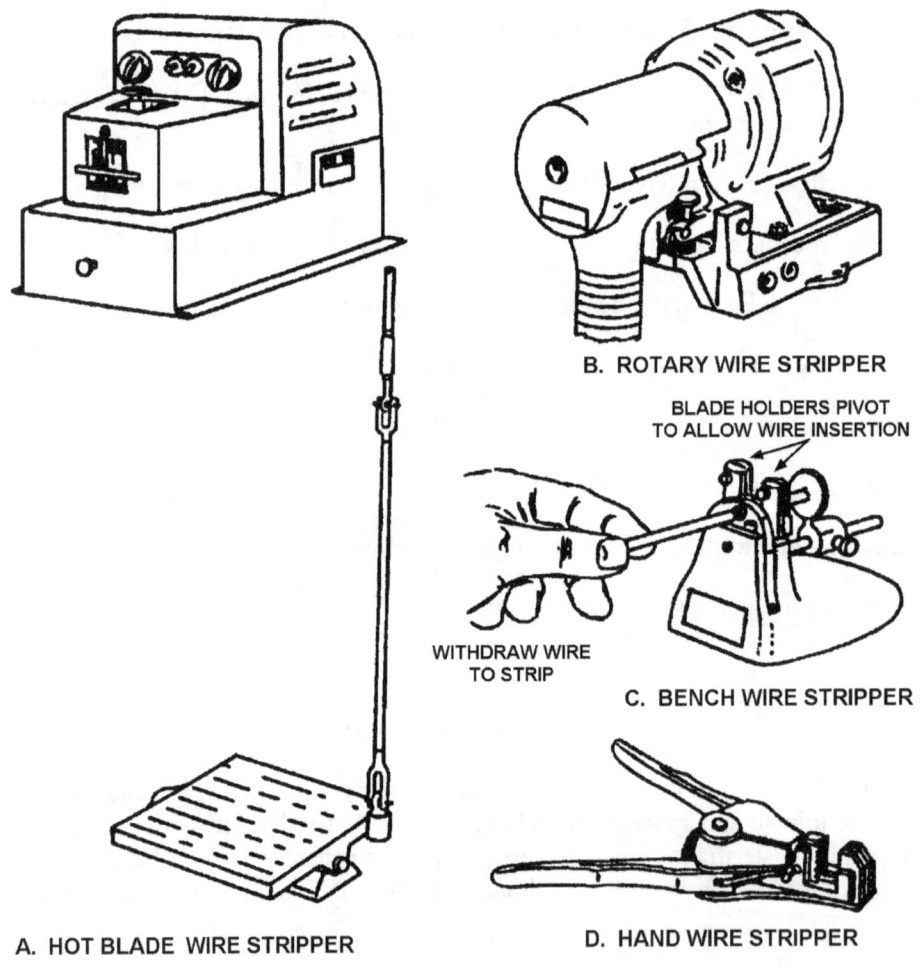

B. ROTARY WIRE STRIPPER

BLADE HOLDERS PIVOT
TO ALLOW WIRE INSERTION

WITHDRAW WIRE
TO STRIP

C. BENCH WIRE STRIPPER

A. HOT BLADE WIRE STRIPPER

D. HAND WIRE STRIPPER

Figure 2-1.—Typical wire-stripping tools.

Hand Wire Stripper

The procedure for stripping wire with the hand wire stripper is as follows (refer to figure 2-2):

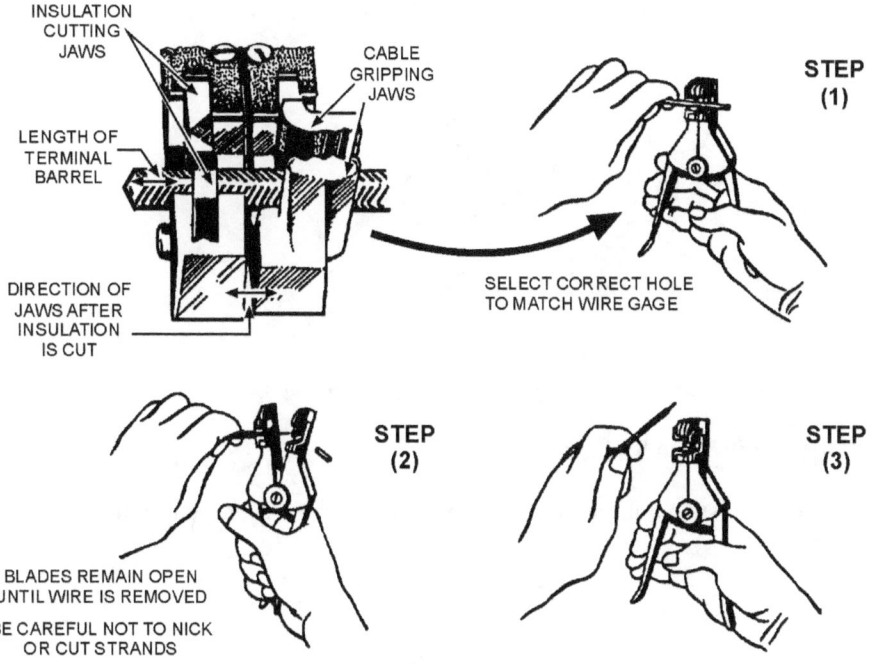

Figure 2-2.—Stripping wire with a hand stripper.

1. Insert the wire into the center of the correct cutting slot for the wire size to be stripped. The wire sizes are listed on the cutting jaws of the hand wire strippers beneath each slot.

2. After inserting the wire into the proper slot, close the handles together as far as they will go.

3. Slowly release the pressure on the handles so as not to allow the cutting blades to make contact with the stripped conductor. On some of the newer style hand wire strippers, the cutting jaws have a safety lock that helps prevent this from happening. Continue to release pressure until the gripper jaws release the stripped wire, then remove.

Knife Stripping

A sharp knife may be used to strip the insulation from a conductor. The procedure is much the same as for sharpening a pencil. The knife should be held at approximately a 60° angle to the conductor. Use extreme care when cutting through the insulation to avoid nicking or cutting the conductor. This procedure produces a taper on the cut insulation as shown in figure 2-3.

WRONG

RIGHT

Figure 2-3.—Knife stripping.

Locally Made Hot-Blade Wire Stripper

If you are required to strip a large number of wires, you can use a locally made hot-blade stripper (figure 2-4) as follows:

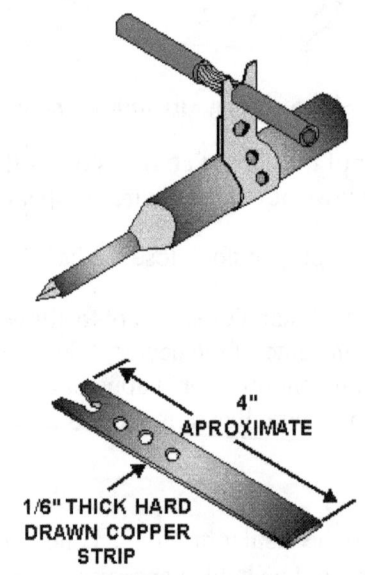

4"
APROXIMATE

1/6" THICK HARD DRAWN COPPER STRIP

Figure 2-4.—Locally made hot-blade stripper.

1. In the end of a piece of copper strip, cut a sharp-edged "V." At the bottom of the "V," make a wire slot of suitable diameter for the size wire to be stripped.

2. Fasten the copper strip around the heating element of an electric soldering iron as shown in figure 2-4. The iron must be rated at 100 watts or greater in order to transfer enough heat to the copper strip to melt the wire insulation.

3. Lay the wire or cable to be stripped in the "V"; a clean channel will be melted in the insulation.

4. Remove the insulation with a slight pull.

General Wire-Stripping Instructions

When stripping wire with any of the tools mentioned, observe the following precautions:

1. Do not attempt to use a hot-blade stripper on wiring with glass braid or asbestos insulation. These insulators are highly heat resistant.

2. When using the hot-blade stripper, make sure the blades are clean. Clean the blades with a brass wire brush as necessary.

3. Make sure all stripping blades are sharp and free from nicks, dents, and so forth.

4. When using any type of wire stripper, hold the wire perpendicular to the cutting blades.

5. Make sure the insulation is clean-cut with no frayed or ragged edges; trim if necessary.

6. Make sure all insulation is removed from the stripped area. Some types of wire are supplied with a transparent layer between the conductor and the primary insulation. If this is present, remove it.

7. When the hand strippers are used to remove lengths of insulation longer than 3/4 inch, the stripping procedure must be done in two or more operations. The strippers will only strip about 3/4 inch at one time.

8. Retwist strands by hand, if necessary, to restore the natural lay and tightness of the strands.

9. Strip aluminum wires with a knife as described earlier. Aluminum wire should be stripped very carefully. Care should be taken not to nick the aluminum wire as the strands break very easily when nicked.

Q1.   What are the basic requirements for any splice or terminal connection?

Q2.   What is the preferred method for stripping wire?

Q3.   What stripping tool would NOT be used to strip glass braid insulation?

Q4.   What tool should be used to strip aluminum wire?

TYPES OF SPLICES

There are six commonly used types of splices. Each has advantages and disadvantages for use. Each splice will be discussed in the following section.

Western Union Splice

The Western Union splice joins small, solid conductors. Figure 2-5 shows the steps in making a Western Union splice.

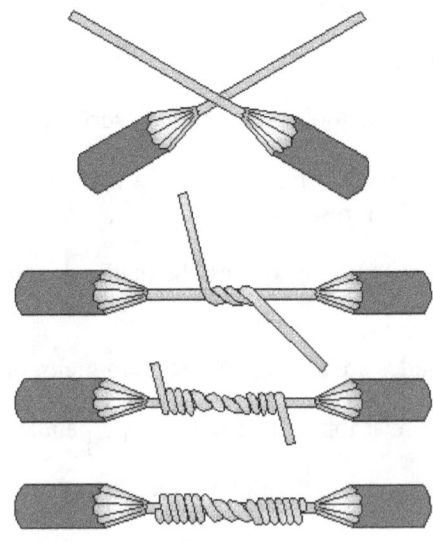

Figure 2-5.—Western Union splice.

1. Prepare the wires for splicing. Enough insulation is removed to make the splice. The conductor is cleaned.

2. Bring the wires to a crossed position and make a long twist or bend in each wire.

3. Wrap one end of the wire and then the other end four or five times around the straight portion of each wire.

4. Press the ends of the wires down as close as possible to the straight portion of the wire. This prevents the sharp ends from puncturing the tape covering that is wrapped over the splice. The various types of tape and their uses are discussed later in this chapter.

Staggering Splices

Joining small multiconductor cables often presents a problem. Each conductor must be spliced and taped. If the splices are directly opposite each other, the overall size of the joint becomes large and bulky. A smoother and less bulky joint can be made by staggering the splices.

Figure 2-6 shows how a two-conductor cable is joined to a similar size cable by using a Western Union splice and by staggering the splices. Care should be taken to ensure that a short wire from one side of the cable is spliced to a long wire, from the other side of the cable. The sharp ends are then clamped firmly down on the conductor. The figure shows a Western Union splice, but other types of splices work just as well.

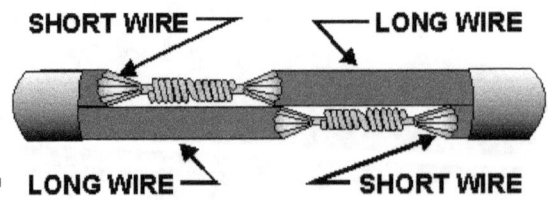

Figure 2-6.—Staggering splices.

Rattail Joint

A splice that is used in a junction box and for connecting branch circuits is the rattail joint (figure 2-7).

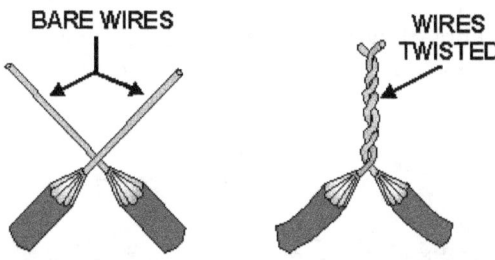

Figure 2-7.—Rattail joint.

Wiring that is installed in buildings is usually placed inside long lengths of steel or aluminum pipe called a conduit. Whenever branch or multiple circuits are needed, junction boxes are used to join the conduit.

To create a rattail joint, first strip the insulation off the ends of the conductors to be joined. You then twist the wires to form the rattail effect. This type of splice will not stand much stress.

Fixture Joint

The fixture joint is used to connect a small-diameter wire, such as in a lighting fixture, to a larger diameter wire used in a branch circuit. Like the rattail joint, the fixture joint will not stand much strain.

Figure 2-8 shows the steps in making a fixture joint. The first step is to remove the insulation and clean the wires to be joined. After the wires are prepared, the fixture wire is wrapped a few times around the branch wire. The end of the branch wire is then bent over the completed turns. The remainder of the bare fixture wire is then wrapped over the bent branch wire. Soldering and taping completes the job.

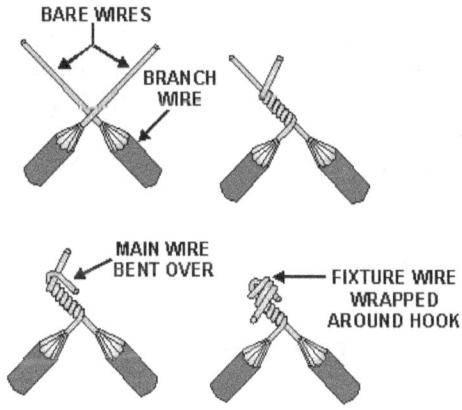

Figure 2-8.—Fixture joint.

Knotted Tap Joint

All the splices discussed up to this point are known as <u>butted</u> splices. Each was made by joining the free ends of the conductors together. Sometimes, however, it is necessary to join a branch conductor to a continuous wire called the main wire. Such a junction is called a tap joint.

The main wire, to which the branch wire is to be tapped, has about 1 inch of insulation removed. The branch wire is stripped of about 3 inches of insulation. The knotted tap is shown in figure 2-9.

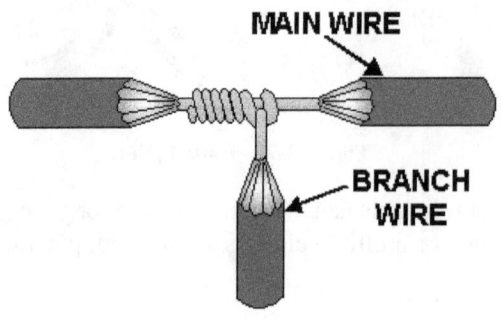

Figure 2-9.—Knotted tap joint.

The branch wire is laid behind the main wire. About three-fourths of the bare portion of the branch wire extends above the main wire. The branch wire is brought under the main wire, around itself, and then over the main wire to form a knot. The branch wire is then wrapped around the main conductor in short, tight turns; and the end is trimmed off.

The knotted tap is used where the splice is subject to strain or slippage. When there is no strain, the knot may be eliminated.

Wire Nut and Split Bolt Splices

The wire nut (view A of figure 2-10) is a device commonly used to replace the rattail joint splice. The wire nut is housed in plastic insulating material. To use the wire nut, place the two stripped conductors into the wire nut and twist the nut. In so doing, this will form a splice like the rattail joint and insulate itself by drawing the wire insulation into the wire nut insulation.

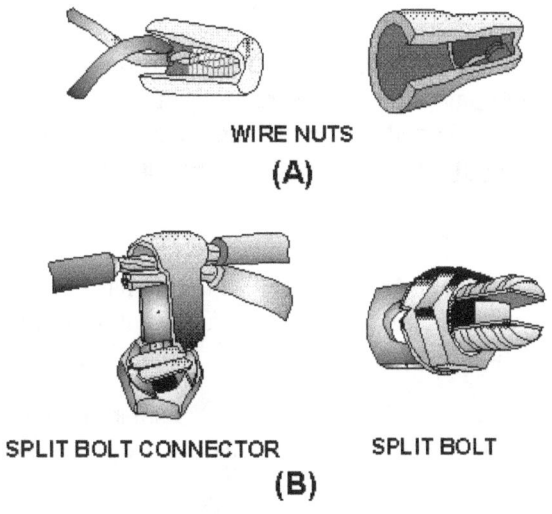

WIRE NUTS
(A)

SPLIT BOLT CONNECTOR   SPLIT BOLT
(B)

Figure 2-10.—Wire nut and split bolt splices.

The split bolt splice (view B of figure 2-10) is used extensively to join large conductors. In the illustration, it is shown replacing the knotted tap joint. The split bolt splice can also be used to replace the "butted" splices mentioned previously when using large conductors.

Q5. Why are the ends of the wire clamped down after a Western Union splice is made?

Q6. Why are splices staggered on multiconductor cables?

Q7. Where is the rattail joint normally used?

Q8. Which type of splice is used to splice a lighting fixture to a branch circuit?

SPLICE INSULATION

The splices we have discussed so far are usually insulated with tape. The following discussion will cover some characteristics of rubber, friction, and plastic insulation tapes.

Rubber Tape

Latex (rubber) tape is a splicing compound. It is used where the original insulation was rubber. The tape is applied to the splice with a light tension so that each layer presses tightly against the one beneath it. This pressure causes the rubber tape to blend into a solid mass. Upon completion, insulation similar to the original is restored.

In roll form, there is a layer of paper or treated cloth between each layer of rubber tape. This layer prevents the latex from fusing while still on the roll. The paper or cloth is peeled off and discarded before the tape is applied to the splice.

The rubber splicing tape should be applied smoothly and under tension so no air space exists between the layers. Start the first layer near the middle of the joint instead of the end. The diameter of the completed insulated joint should be somewhat greater than the overall diameter of the original wire, including the insulation.

WARNING

Some rubber tapes are made for special applications. These types are semiconducting and will pass electrical current, which presents a shock hazard. These types of tape are packaged similar to the latex rubber tape. Care should be taken to insulate splices <u>only</u> with latex rubber insulating tape.

Friction Tape

Putting rubber tape over the splice means that the insulation has been restored to a great degree. It is also necessary to restore the protective covering. Friction tape is used for this purpose. It also provides a minor degree of electrical insulation.

Friction tape is a cotton cloth that has been treated with a sticky rubber compound. It comes in rolls similar to rubber tape except that no paper or cloth separator is used. Friction tape is applied like rubber tape; however, it does not stretch.

The friction tape should be started slightly back on the original insulation. Wind the tape so that each turn overlaps the one before it. Extend the tape over onto the insulation at the other end of the splice. From this point, a second layer is wound back along the splice until the original starting point is reached. Cutting the tape and firmly pressing down the ends completes the job. When proper care is taken, the splice and insulation can take as much abuse as the rest of the original wire.

Plastic Electrical Tape

Plastic electrical tape has come into wide use in recent years. It has certain advantages over rubber and friction tape. For example, it can withstand higher voltages for a given thickness. Single thin layers of certain plastic tape will withstand several thousand volts without breaking down. However, to provide an extra margin of safety, several layers are usually wound over the splice. The extra layers of thin tape add very little bulk. The additional layers of plastic tape provide the added protection normally furnished by friction tape.

Plastic electrical tape usually has a certain amount of stretch so that it easily conforms to the contour of the splice.

Q9. *Which of the splices discussed is NOT a butted splice?*

Q10. *Why is friction tape used in splicing?*

TERMINAL LUGS

Since most cable wires are stranded, it is necessary to use terminal lugs to hold the strands together to aid in fastening the wires to terminal studs (see figure 2-11). The terminals used in electrical wiring are either of the soldered or crimped type. Terminals used in repair work must be of the size and type specified on the electrical wiring diagram for the particular equipment.

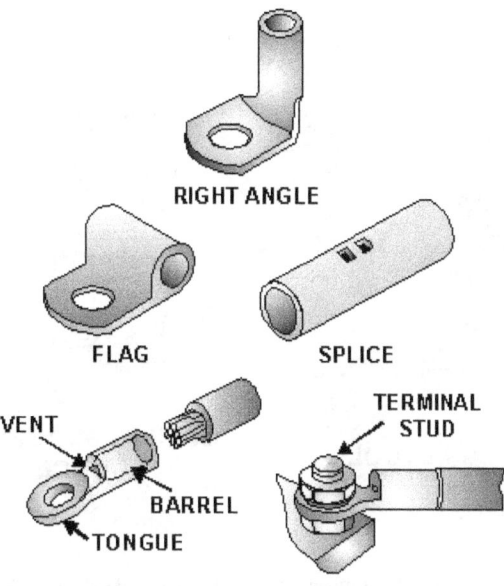

Figure 2-11.—Noninsulated terminal lugs and splices.

The increased use of crimp-on terminals is due to the limitations of soldered terminals. The quality of soldered connections depends mostly upon the operator's skill. Other factors, such as temperature, flux, cleanliness, oxides, and insulation damage due to heat, also add to defective connections. Solder-type connections are covered later in this chapter.

An advantage of the crimp-on solderless terminal lugs is that they require relatively little operator skill to use. Another advantage is that the only tool needed is the crimping tool. This allows terminal lugs to be applied with a minimum of time and effort. The connections are made rapidly, are clean, and uniform in construction. Because of the pressures exerted and the material used, the crimped connection or splice, properly made, is both mechanically and electrically sound. Some of the basic types of terminals are shown in figure 2-11. There are several variations of these basic types, such as the use of a slot instead of a terminal hole, three- and four-way splice-type connectors, and others.

Since the Navy uses both copper and aluminum wiring, both copper and aluminum terminals are necessary. Various size terminal or stud holes may be found for each of the different wire sizes. A further refinement of the solderless terminals and splices is the insulated type. The barrel of the terminal or splice is enclosed in an insulated material. The insulation is compressed along with the terminal barrel when it is crimped, but is not damaged in the process. This rids you of the need for taping or tying an insulating sleeve over the joint.

There are several different types of crimping tools used with copper terminals. However, you will normally be concerned only with wire sizes AWG (American Wire Gauge) 10 or smaller. For wire of these sizes, a small plier-type crimper is used to crimp on uninsulated terminals, as shown in figure 2-12. The small plier-type crimper has several sizes of notches for the different size terminals. Care should be used to select the correct size crimping tool for the particular terminal.

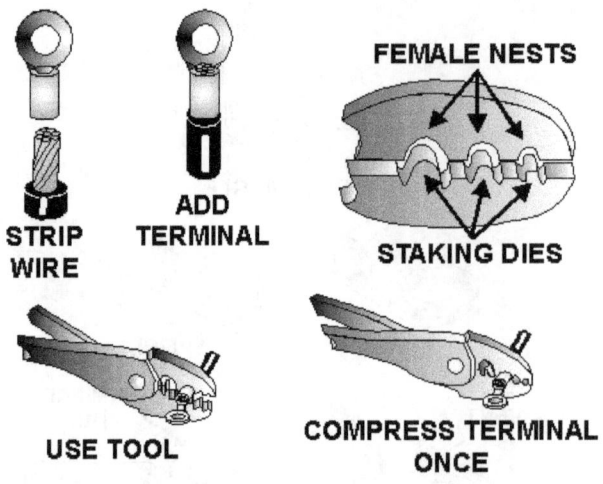

Figure 2-12.—Crimping small copper uninsulated terminals.

NONINSULATED TERMINAL AND SPLICE INSULATION

When noninsulated terminals and splices are used, some form of insulation must be used to cover the bare conductor. The two most common forms of insulator used for terminals and splices are transparent tubing (commonly called spaghetti) and heat-shrinkable tubing. If spaghetti is used, it must be tied with lacing twine, as illustrated in figure 2-13. Heat-shrinkable tubing is shrunk to the desirable size by applying dry heat. It is also a good way to insulate terminals and splices, as illustrated in figure 2-14. This tubing shrinks to approximately one-half its original diameter when heated with an electrical hot-air gun (figure 2-15). Here are the steps for using the hot-air gun:

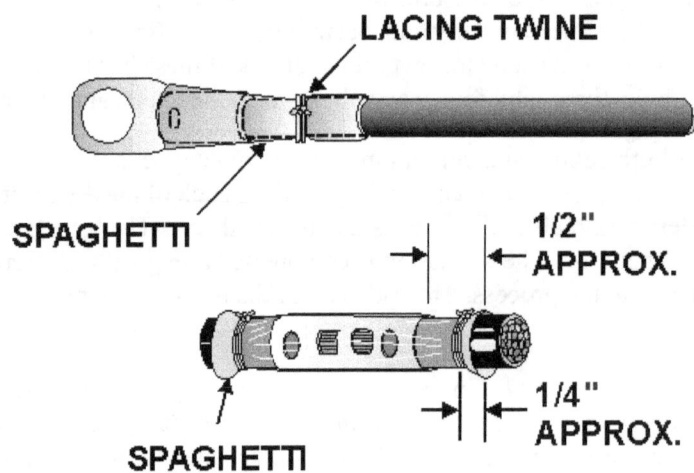

Figure 2-13.—Spaghetti tied with lacing twine.

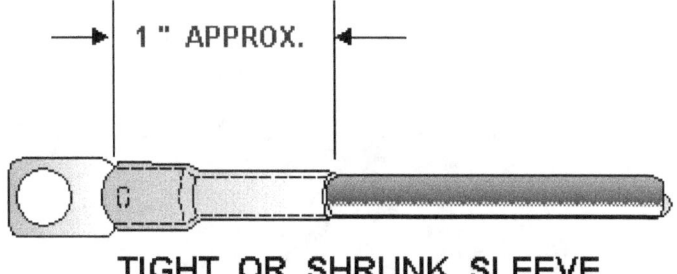

TIGHT OR SHRUNK SLEEVE

Figure 2-14.—Shrunken sleeve.

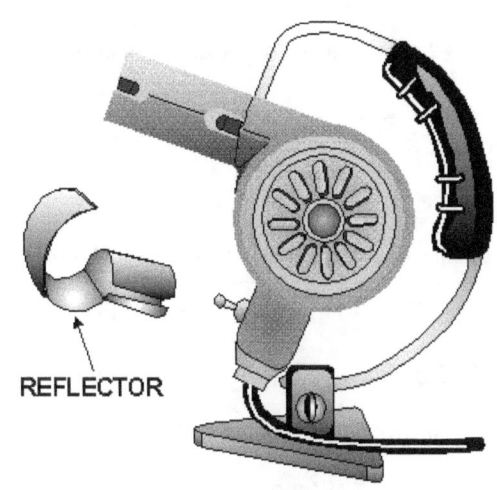

REFLECTOR

Figure 2-15.—Typical hot-air gun.

1. Hold the heat source 4 to 5 inches away from the wire. Apply a heat of 275° F to 300° F for about 30 seconds. Rotate the wire while applying the heat so that the heat is evenly distributed.

2. Remove the heat as soon as the tubing conforms to the shape of the wire. Allow the tubing to cool for at least 30 seconds before handling.

CAUTION

Do not apply heat higher than 300° F as this may damage the wire. Do not continue to apply heat after the tubing has shrunk onto the wire. Further application of heat will not cause additional shrinkage of the tubing.

COMPRESSED AIR/NITROGEN HEATING TOOL

The compressed air/nitrogen heating tool (figure 2-16) is a new tool in the fleet and was designed as a portable source of heat. This tool is safe for use around fueled aircraft because an open heating element is not required. The compressed air/nitrogen heating tool can be used on heat-shrinkable tubing.

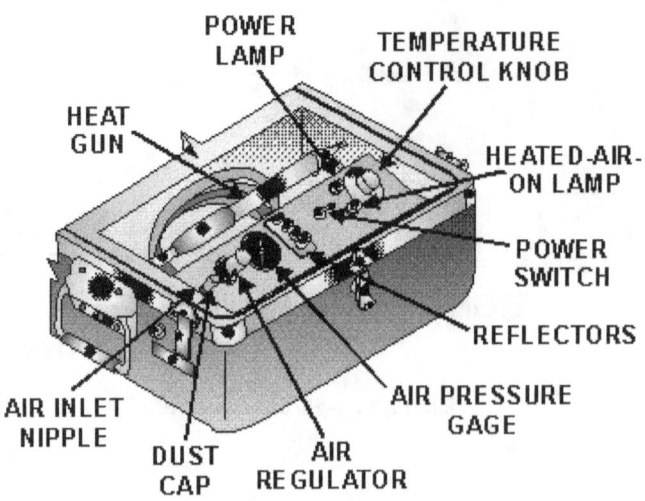

Figure 2-16.—Compressed air/nitrogen heating tool.

The compressed air/nitrogen heating tool comes in two styles: ac or dc electrical power supplies. The power requirements are listed in table 2-1 for both styles.

Table 2-1.—Compressed Air/Nitrogen Heating Tool Power Requirements

Electrical Power, HT-900B	115 VAC, 50-400 Hz, single-phase, 7 Amps
Electrical Power, HT-920B	220 VAC, 50-400 Hz, single-phase, 3.5 Amps
Heat gun output temperature	550-920°F (290-495°C)
Compressed air/nitrogen	80-200 psig, 4 SCFM (Dry and oil-free)

Refer to the operator's manual for safe operating procedures for the compressed air/nitrogen heating tool. A brief summary of these procedures follows:

1. Push down and fully turn the air regulator knob counterclockwise. This is to ensure that the air regulator is off.

2. Remove the dust cap from the air inlet nipple. The inlet nipple is what we connect the air or nitrogen source line to.

WARNING

If nitrogen is used, make sure that you are in a well-ventilated area. Using nitrogen is a poorly ventilated area. Using nitrogen is a poorly ventilated area can result in suffocation.

CAUTION

As noted in table 2-1, the compressed air/nitrogen source CANNOT be greater than 200 psig.

3. Attach the air/nitrogen hose to the inlet nipple, making sure there is a firm connection.

4. Once the air/nitrogen source is properly attached, push down and turn the air regulator knob clockwise until the pressure on the air pressure gauge indicates between 5 to 7 psig.

5. Plug in the power cord to an appropriate grounded power supply.

6. Set the power switch to the ON position. The power lamp and heated-air-on lamp will both illuminate. (If the lights do not come on, check the switch on the gun handle. The switch must be positioned toward the front of the handle.)

7. There is a 1-minute warm-up time. During this warm-up period, ensure that the indicated air pressure increases to 10 to 15 psig on the air-pressure gauge.

8. You can now adjust the temperature control knob to the desired temperature setting.

9. You can turn the air/nitrogen pressure off and on to the gun without powering down the module by using the switch mounted on the gun handle.

After you complete your task with the compressed air/nitrogen heating tool, use the following shutdown procedures:

1. Push down and fully turn the air regulator knob counterclockwise. Observe that the air pressure gauge indication drops to 0 psig and the heated air lamp goes out.

2. Position the switch on the heating gun toward the rear of the handle.

3. Place the power switch to the OFF position and observe that the power lamp goes out.

4. Allow the air/nitrogen to flow for a minimum of 1 minute to cool the heating gun. (This procedure is done to extend the life of the heating element.)

5. Disconnect the power connecter from the power source.

6. Turn off air/nitrogen source at place of origin and disconnect.

7. Disconnect the compressed air/nitrogen hose from the air inlet nipple and install the dust cap on the air inlet nipple.

Noninsulated Copper Terminals

The procedure for crimping a copper terminal (noninsulated) to a copper wire is as follows:

1. With a wire stripper, trim the insulation from the wire about one thirty-second of an inch longer than the length of the terminal barrel. When using a wire stripper, be sure to use the correct size stripping slot for the wire size used. Otherwise, all the insulation will not be removed or, if the slot is too small, the outside strands of the conductor will be nicked and consequently weakened. When a knife is used for stripping wire, care should be used to prevent nicking the strands. Slip the spaghetti or heat-shrinkable tubing over the wire and back far enough to be out of the way of the crimping operation.

2. Slip the terminal barrel over the bared wire end and up against the insulation. Make certain that all wire strands are inside the tubular barrel of the terminal.

3. Center the terminal barrel in the female nest of the plier jaws as shown in figure 2-12 so that the indentation formed by the staking die will be in the center of the barrel. Crimp until the pliers reach their stop or limit. This is necessary for a good mechanical and electrical connection.

4. Slip the tubular insulation down over the terminal barrel so that it extends a little beyond the barrel. Tie it in place if spaghetti is used. If heat-shrinkable tubing is used, shrink with a heat gun.

Q11. *What is a major advantage of the crimped terminal over the soldered terminal?*

Q12. *What are the two types of insulation most commonly used for noninsulated splices and terminal lugs?*

Q13. *What is the maximum allowable temperature that should be used on heat-shrinkable tubing?*

Q14. *What is the maximum allowable source pressure that can be used with the compressor air/nitrogen heating tool?*

ALUMINUM TERMINALS AND SPLICES

Terminals that are used with aluminum wire are made of aluminum. Proper crimping is more difficult with these terminals because of such factors as aluminum creep and softness. Aluminum wire has an undesirable characteristic called aluminum creep. Aluminum has the tendency to actually move away from the point where pressure is applied. This is not only true during the crimping operation but also takes place during temperature changes. The aluminum wire is softer than the terminal lugs and splice connectors and contracts faster than the connector when the temperature drops. This causes the wires to creep away from the crimped connections, which, in turn, causes loose connections. The softness of aluminum wire also makes it subject to being cut or nicked during stripping. You should be careful never to use an aluminum terminal with copper wire or a copper terminal with aluminum wire because of electrolysis. Electrolysis is the chemical action that takes place when an electric current passes through two dissimilar metals. This chemical action corrodes (eats away) the metal. Also, never use the aluminum crimping tool for crimping other than the aluminum terminals. Aluminum terminal lugs and splices are not insulated, so you must use spaghetti or heat-shrinkable tubing for insulation as discussed earlier.

The barrels of several styles of larger size aluminum terminal lugs are filled with a petroleum abrasive compound. This compound causes a grinding action during the crimping operation. This removes the oxide film from the aluminum. It also prevents the oxide film from reforming in the connection. All aluminum terminals and splices have an inspection hole to allow checking the depth of wire insertion. This hole is sealed with a removable plug, which also serves to hold in the oxide-inhibiting compound (figure 2-17).

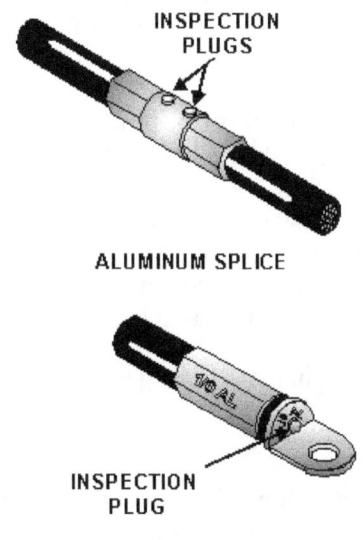

ALUMINUM SPLICE

ALUMINUM TERMINAL LUG

Figure 2-17.—Aluminum terminal lug and splice.

It is recommended that only power-operated crimping tools be used to install large aluminum terminal lugs and splices. (See view A of figure 2-18.)

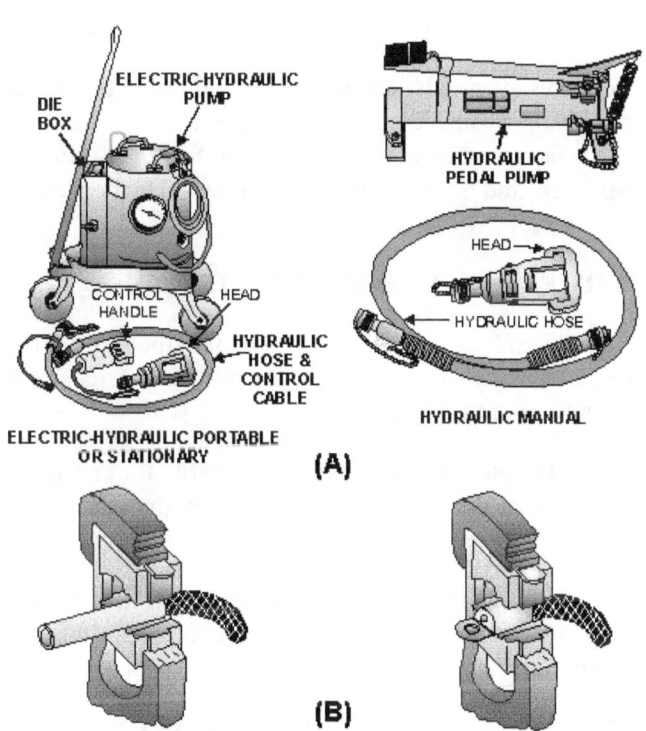

Figure 2-18.—Power crimping tools.

The steps used for crimping an aluminum terminal or splice to an aluminum wire (view B of figure 2-18) are as follows:

1. Carefully remove the conductor insulation. Do not cut or nick the aluminum conductors. Do not wire-brush or scrape the aluminum conductor (the compound in the terminal or splice barrel will clean it satisfactorily).

2. Remove the protective foil wrapping from the terminal or splice and check the amount of compound in the terminal barrel. It should be one-fourth to one-half full.

3. Slip the spaghetti or heat-shrinkable tubing over the wire and back far enough to be out of the way of the crimping operation. Insert the stripped conductor the full length of the terminal or splice barrel. While doing this, leave the plug over the inspection hole. This allows the compound to be forced in and around the strands.

4. Center the terminal lug or splice in the crimping tool.

5. Actuate the power crimping tool.

6. Wipe off the excess compound. Inspect the joint with a probe through the inspection hole. The end of the conductor should come to the edge of the inspection hole.

7. Slip the tubular insulation down over the terminal or splice barrel. Tie it in place if spaghetti is used. If using heat-shrinkable tubing, shrink with a heat gun.

Q15.   Should aluminum wire be cleaned prior to installing an aluminum terminal lug or splice?

Q16.   What tools should be used to install large aluminum terminal lugs and splices?

Q17.   Why should a lockwasher never be used with an aluminum terminal?

Improper crimping procedures eventually cause terminal failure. Be especially careful of undercrimping, overcrimping, using wrong crimping tools, improper cleaning methods, and cutting or nicking the conductors. A loose contact allows an oxide film to form between the wire and the terminal. This results in increased resistance, and the resistance causes heat. The heat accelerates deterioration, and eventually a failure results.

PREINSULATED COPPER TERMINAL LUGS AND SPLICES

The use of preinsulated terminal lugs and splices has become the most common method for copper wire termination and splicing in recent years. It is by far the best and easiest method. There are many tools used for crimping terminal lugs and splices.

Hand, portable power, and stationary power tools are available for crimping terminal lugs. These tools crimp the barrel to the conductor and, at the same time, form the insulation support to the wire insulation.

The power tools, both stationary and portable, are usually found in large shops where wire bundles are made up. In the next paragraphs, we will discuss the more common hand-crimping tools you will most likely be using in your day-to-day work.

TERMINATING COPPER WIRE WITH PREINSULATED TERMINAL LUGS

Small-diameter copper wires are terminated with solderless, preinsulated copper terminal lugs. As shown in figure 2-19, the insulation is part of the terminal lug. It extends beyond the barrel so that it covers a portion of the wire insulation. This makes the use of spaghetti or heat-shrinkable tubing unnecessary. Preinsulated terminal lugs also have an insulation support (a metal reinforcing sleeve) beneath the insulation for extra supporting strength of the wire insulation. Some preinsulated terminals fit more than one size of wire. The insulation is color coded, and the range of wire sizes is marked on the tongue. This identifies the wire sizes that can be terminated with each of the terminal lug sizes. (See table 2-2.)

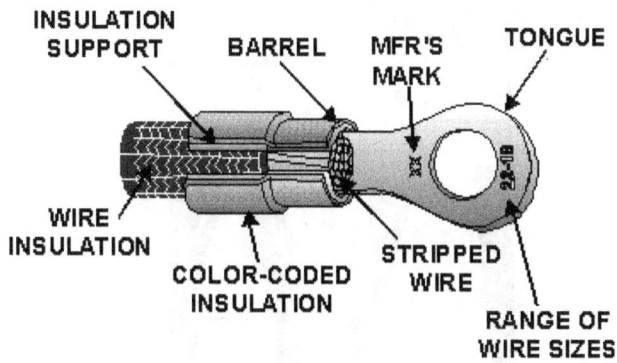

Figure 2-19.—Preinsulated straight copper terminal lug.

Table 2-2.—Color Coding of Copper Terminal Lug or Splice Insulation

Color of Terminal Lug or Splice Insulation	To Be Used on Wire Sizes
Yellow (Bright)	#26 - #24
Red	#22 - #20, #18
Blue	#16 - #14
Yellow (Dull)	#12 - #10

For crimping small copper terminal lugs, several hand-crimping tools can be used for wire sizes AWG 26 through 10 (figure 2-20). These hand-crimping tools have a self-locking ratchet, which prevents the tool from opening until the crimp is completed. Some of these tools have a color-coded selector knob to match the color-coded terminal lug or splice being used. Other tools have a replaceable set of dies for several wire sizes. The hand-crimping procedure for preinsulated copper terminal lugs in wire sizes No. 26 through No. 10 with the standard hand-crimp tool is as follows:

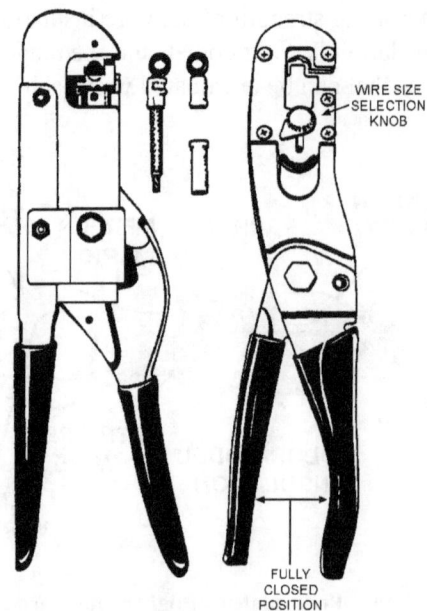

Figure 2-20.—Hand-crimping tools.

1. Strip the wire insulation using the recommended stripping procedures already discussed.

2. Ensure that the tool handles are fully open and the proper die set has been installed correctly.

3. Insert the terminal lug, tongue first, into the wire side of the hand tool barrel crimping jaws. Be certain the terminal lug barrel butts flush against the tool stop on the locator. See figure 2-21 for the correct insertion method.

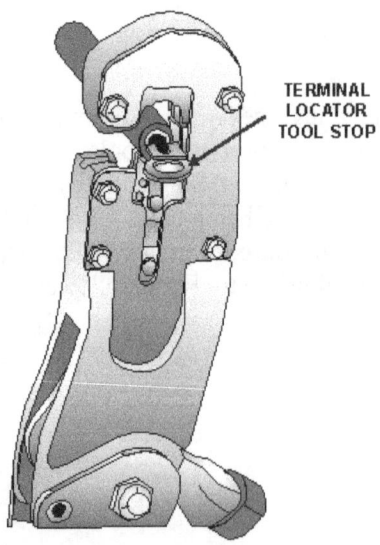

Figure 2-21.—Crimping tool with terminal lug inserted.

4. Squeeze the tool handles slowly until the tool jaws hold the terminal lug barrel firmly in place, but without denting it.

5. Insert the stripped wire into the terminal lug barrel until the wire insulation butts flush against the near end of the wire barrel. (See figure 2-22.)

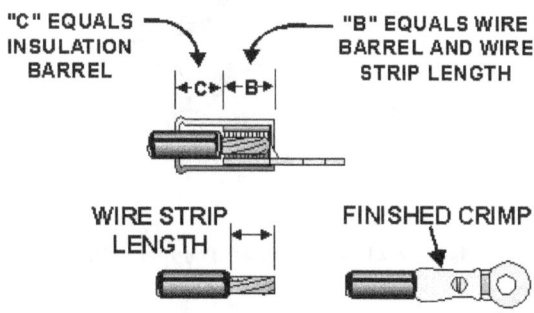

Figure 2-22.—Proper insertion of stripped wire in insulation terminal lug for crimping.

6. Squeeze the tool handles until the rachet releases.

7. Remove the completed assembly and examine it for the proper crimp in accordance with the following:

    a. Indent centered on the terminal lug barrel.

    b. Indent in line with the barrel.

    c. Terminal lug not cracked.

    d. Terminal lug insulation not cracked.

    e. Insulation grip crimped.

CAUTION

If not properly stripped, some of the smaller gauge, thin-wall wire insulation can be inadvertently inserted and crimped in the terminal wire barrels. This will cause a bad electrical connection. Do not use any connection that is found defective as a result of a visual inspection. Cut off the defective connection and remake using a new terminal lug.

PREINSULATED SPLICES

Preinsulated permanent copper splices are used to join small copper wire AWG sizes No. 26 through No. 10. A typical splice is shown in figure 2-23. Note that the splice preinsulation extends over the wire insulation. Each splice size can be used for more than one wire size. Splices are color coded in the same manner as preinsulated small copper terminal lugs (see table 2-2).

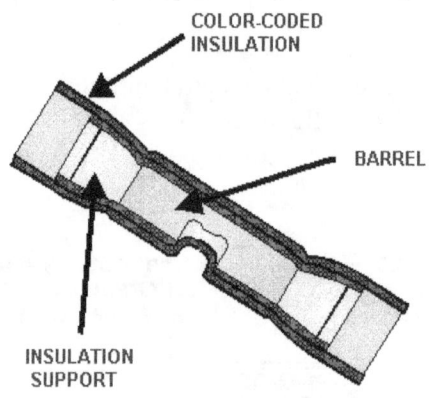

Figure 2-23.—Preinsulated copper splice.

Crimping Procedure for Splices.

Crimping small preinsulated copper splices in the No. 26 to No. 14 wire-size range can be accomplished with several recommended tools. In this section, we will discuss the basic crimping procedures.

1. Strip wire to length following one of the procedures already discussed.

2. With the tool handles fully open, set the wire size selector knob to the proper position for the wire size being crimped. Slide the terminal lug locator down below the die surface into the fully retracted position. (See figure 2-24.) Slide the splice locator back into the retracted position. Insert the splice into the tool so that the "locating shoulder" on the side of the splice to be crimped is in the space between the two crimping dies. The insulation barrel on this side of the splice should protrude from the "wire side" of the tool. (See figure 2-24.) Slide the splice locator into the fully extended position. Insert the splice into the stationary die so that the locator "finger" fits into the locator groove in the splice.

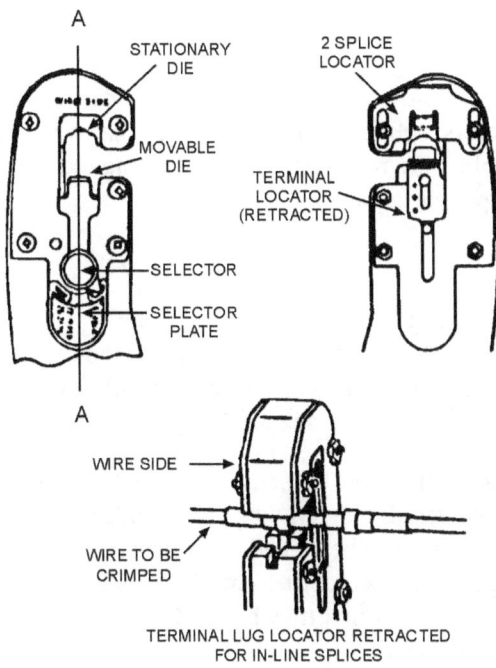

Figure 2-24.—Locating splice in crimping tool.

3. Squeeze the tool handles slowly until the tool jaws hold the spice barrel firmly in place, but without denting he barrel.

4. Insert the stripped wire into the splice barrel, which protrudes from the "wire side" of the splice, until the stripped end of wire butts against the stop in the center of the splice. This can be seen through the splice inspection window.

5. Crimp by closing the tool handles. The tool will not open until the full crimping cycle has been completed.

6. After crimping, check that the wire end is still visible through the splice inspection window.

7. Reverse the position of the splice in the crimping tool (or location of the crimping tool on the splice) and repeat steps 1 through 6 to crimp the wire into the other side of the splice.

If the correct tools are used and the proper procedures followed, crimp-on connections are more effective electrically, as well as mechanically, than soldered connections. A visual inspection is very important. It reveals oxidation, deterioration, overheating, and broken conductors. In some cases it may be necessary to check these connections with an ohmmeter. The proper resistance, for all practical purposes, should be zero. Any defective terminal should be removed and a new terminal crimped on.

Q18. *What is the most common method of terminating and splicing wires?*

Q19. *Besides not having to insulate a noninsulated terminal, what other advantage is gained by using a preinsulated terminal lug?*

Q20. *Why are preinsulated terminal lugs and splices color coded?*

SOLDERING

The following information will aid you in learning basic soldering skills. It should enable you to solder wires to electrical connectors, splices, and terminal lugs that we have discussed earlier in the chapter. Special skills and schooling are required for the soldering techniques used in printed circuit boards and microminiature component repair.

SOLDERING PROCESS

Cleanliness is essential for efficient, effective soldering. Solder will not adhere to dirty, greasy, or oxidized surfaces. Heated metals tend to oxidize rapidly. This is the reason the oxides, scale, and dirt must be removed by chemical or mechanical means. Grease or oil films can be removed with a suitable solvent. Connections to be soldered should be cleaned just prior to the actual soldering operation.

Items to be soldered should normally be "tinned" before making a mechanical connection. Tinning is the coating of the material to be soldered with a light coat of solder. When the surface has been properly cleaned, a thin, even coating of flux should be placed over the surface to be tinned. This will prevent oxidation while the part is being heated to soldering temperature. Rosin-core solder is usually preferred in electrical work. However, a separate rosin flux may be used instead. Separate rosin flux is frequently used when wires in cable fabrication are tinned.

Q21.   Why must items to be soldered be cleaned just prior to the soldering process?

TINNING COPPER WIRE AND CABLE

Wires to be soldered to connectors should be stripped so that when the wire is placed in the barrel, there will be a gap of approximately 1/32 inch between the end of the barrel and the end of the insulation. This is done to prevent burning the insulation during the soldering process and to allow the wire to flex easier at a stress point. Before copper wires are soldered to connectors, the ends exposed by stripping are tinned to hold the strands solidly together. The tinning operation is satisfactory when the ends and sides of the wire strands are fused together with a coat of solder. Do not tin wires that are to be crimped to solderless terminals or splices.

Copper wires are usually tinned by dipping them into flux (view A of figure 2-25) and then into a solder bath (pot) (view B of the figure). In the field, copper wires can be tinned with a soldering iron and rosin-core solder. Tin the conductor for about half its exposed length. Tinning or solder on the wire above the barrel causes the wire to be stiff at the point where flexing takes place. This will result in the wire breaking.

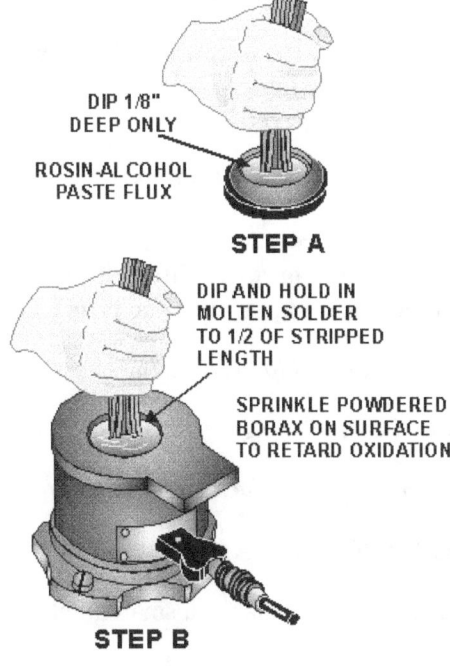

DIP 1/8"
DEEP ONLY

ROSIN-ALCOHOL
PASTE FLUX

STEP A

DIP AND HOLD IN
MOLTEN SOLDER
TO 1/2 OF STRIPPED
LENGTH

SPRINKLE POWDERED
BORAX ON SURFACE
TO RETARD OXIDATION

STEP B

Figure 2-25.—Dip-tinning In a solder pot.

The flux used in tinning copper wire is a mixture of denatured alcohol and freshly ground rosin. This type of flux may be mixed just prior to use. A premixed paste flux may also be used. The solder used for terminal lugs, splices, and connectors is a mixture of 60-percent tin and 40-percent lead. Maintain the temperature of the solder bath (pot) between 450 and 500° F. This keeps the solder in a liquid state. Skim the surface of the solder pot, as necessary, with a metal spoon or blade. This keeps the solder clean and free from oxides, dirt, and so forth.

Dip-tin wires smaller than No. 8 in groups of 8 or 10. Dip-tin wires size No. 8 and larger individually. The procedure for dip-tinning is as follows:

1. Prepare the flux and solder as previously described.

2. Make sure the exposed end of the wire is clean and free from oil, grease, and dirt. Strands should be straight and parallel. Dirty wire should be restripped.

3. Grasp the wire firmly and dip it into the prepared flux to a depth of about 1/8 inch (see view A of figure 2-25).

4. Remove the wire and shake off the excess flux.

5. Immediately dip the wire into molten solder. Dip only half of the stripped conductor length into the solder (see view B of figure 2-25).

6. Turn the wire slowly in the solder bath until the wire is well tinned. Watch the solder fuse to the wire. Do not keep the wire in the bath longer than necessary.

7. Remove the excess solder by wiping the tinned conductor on a cloth.

WARNING

Do not shake off excess solder. It can cause serious burns if it contacts your skin. It can also cause short circuits in exposed electrical equipment that may be in the immediate area of the tinning operation.

CAUTION

Use only rosin flux or rosin-core solder for tinning copper wires to be used in electrical and electronics systems. Corrosive flux will cause damage. During the tinning operation, do not melt, scorch, or burn the insulation.

Q22. What does "tinning" mean in relationship to soldering?

Q23. Why should wire be stripped 1/32 inch longer than the depth of the solder barrel?

Q24. How much of the stripped length of a conductor should be tinned?

ALTERNATIVE DIP-TINNING PROCEDURE

If an electrically heated solder pot is not available, a small number of wires can be tinned using the following procedure (see figure 2-26):

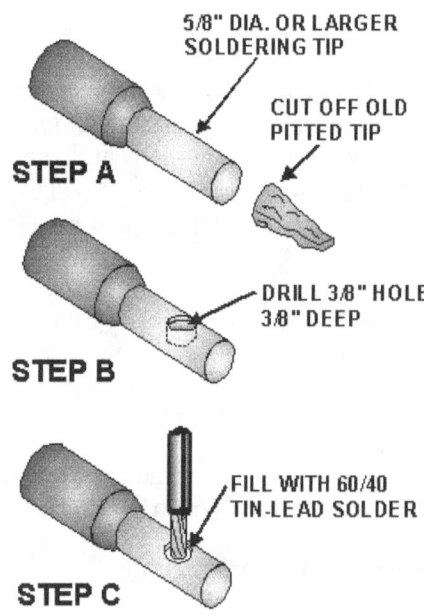

Figure 2-26.—Alternate dip-tinning method.

1. Cut off the beveled section of the tip of a discarded soldering iron tip.

2. Drill a hole (1/4- to 3/8-inch diameter) in the round part of the tip about two-thirds through.

3. Heat the iron and melt the rosin-core solder into the hole.

4. Tin the wires by dipping them into the molten solder one at a time.

5. Keep adding fresh rosin-core solder as the flux burns away.

PROCEDURE FOR TINNING COPPER WIRE WITH A SOLDERING IRON

In the field, wires smaller than size No. 10 can be tinned with a soldering iron and rosin-core solder as follows (see figure 2-27):

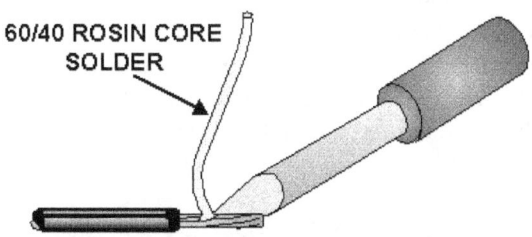

Figure 2-27.—Tinning wire with a soldering iron.

1. Select a soldering iron with the correct heat capacity for the wire size (see table 2-3). Make sure that the iron is clean and well tinned.

Table 2-3.—Approximate Soldering Iron Size for Tinning

Wire Size (AWG)	Soldering Iron Size (Heat Capacity)
#20 - #16	65 Watts
#14 & #12	100 Watts
#10 & #8	20 Watts

2. Start by holding the iron tip and solder together on the wire until the solder begins to flow.

3. Move the soldering iron to the opposite side of the wire and tin half of the exposed length of the conductor.

The tinned surfaces to be joined should be shaped, fitted, and then mechanically joined to make a good mechanical and electrical contact. The parts must be held still. Any motion between the parts while the solder is cooling usually results in a poor solder connection, commonly called a "fractured solder" joint.

Q25. What causes a "fractured solder" joint?

SOLDERING TOOLS

Many types of soldering tools are in use today. Some of the more common types are the soldering iron, soldering gun, resistance soldering set, and pencil iron. The following discussion will provide you with a working knowledge of these tools.

Soldering Irons

Some common types of hand soldering irons are shown in figure 2-28. All high-quality soldering irons operate in the temperature range of 500 to 600° F. Even the 25-watt midget irons produce this temperature. The important difference in iron sizes is not temperature, but thermal inertia. Thermal inertia is the capacity of the iron to generate and maintain a satisfactory soldering temperature while giving up heat to the joint to be soldered. Although it is not practical to solder large conductors with the 25-watt iron, this iron is quite suitable for replacing a half-watt resistor in an electronic circuit or soldering a miniature connector. One advantage of using a small iron for small work is that it is light and easy to handle and has a small tip that is easily used in close places. Even though its temperature is high enough, a midget iron does not have the thermal inertia to solder large conductors.

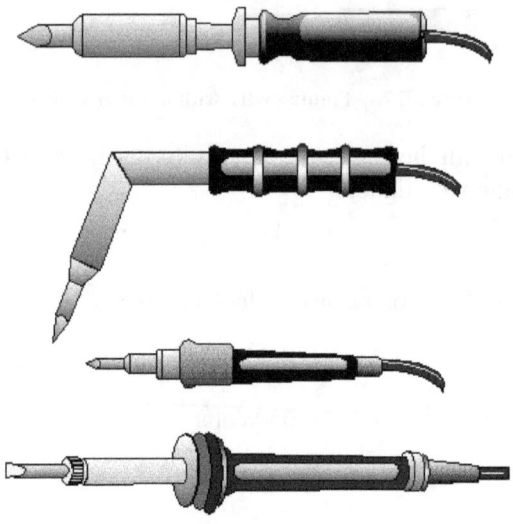

Figure 2-28.—Types of hand soldering Irons.

A well-designed iron is self-regulating. The resistance of its element increases with rising temperature. This limits the flow of current. Some common tip shapes of the soldering irons in use in the Navy are shown in figure 2-29.

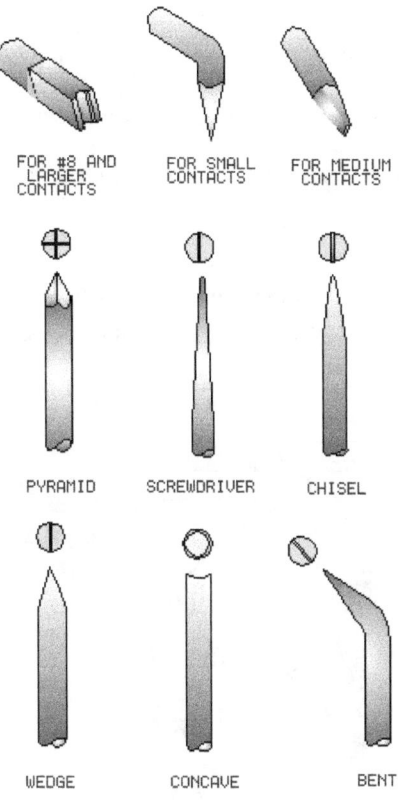

FOR #8 AND LARGER CONTACTS   FOR SMALL CONTACTS   FOR MEDIUM CONTACTS

PYRAMID   SCREWDRIVER   CHISEL

WEDGE   CONCAVE   BENT

Figure 2-29.—Soldering iron tip shapes.

An iron should be tinned (the application of solder to the tip after the iron is heated) prior to soldering a component in a circuit. After extended use of an iron, the tip tends to become pitted due to oxidation. Pitting indicates the need for retinning. The tip is retinned after first filing the tip until it is smooth (see figure 2-30).

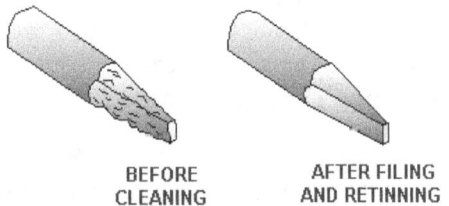

BEFORE CLEANING   AFTER FILING AND RETINNING

SOLDERING IRON TIP BEFORE AND AFTER CLEANING

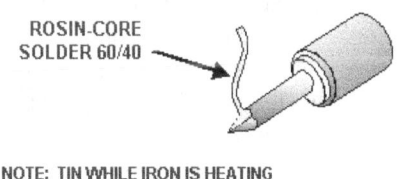

ROSIN-CORE SOLDER 60/40

NOTE: TIN WHILE IRON IS HEATING
TINNING SOLDERING IRON TIP.

Figure 2-30.—Reconditioning pitted soldering iron tip.

Q26. *Define thermal inertia.*

Q27. *Why are small-wattage soldering irons not used to solder large conductors?*

Q28. *State why a well-designed soldering iron is self-regulating.*

Q29. *What should be done to a soldering iron tip that is pitted?*

Soldering Gun

The soldering gun (figure 2-31) has gained great popularity in recent years because it heats and cools rapidly. It is especially well adapted to maintenance and troubleshooting work where only a small part of the technician's time is spent actually soldering.

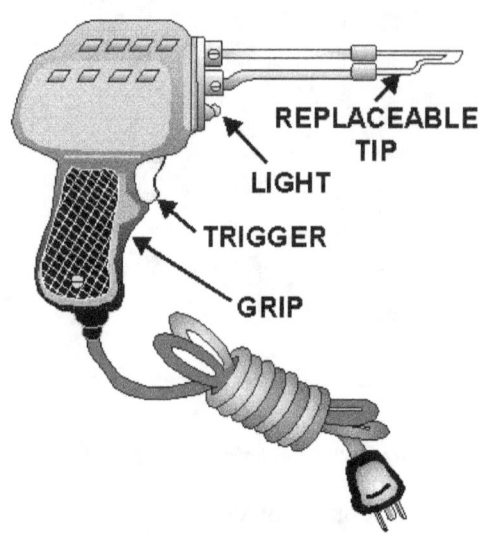

Figure 2-31.—Soldering gun.

A transformer in the soldering gun supplies approximately 1 volt at high current to a loop of copper, which acts as the soldering tip. It heats to soldering temperature in 3 to 5 seconds. However, it may overheat to the point of incandescence if left on over 30 seconds. This should be avoided because excess heat will burn the insulation off the wiring. The gun is operated by a finger switch. The gun heats only while the switch is pressed.

Since the gun normally operates only for short periods at a time, it is comparatively easy to keep clean and well tinned. Short operating time allows little oxidation to form. Because the tip is made of pure copper, it is likely to pit, due to the dissolving action of the solder.

The gun or iron should always be kept tinned to permit proper heat transfer to the connection to be soldered. Tinning also helps control the heat to prevent solder buildup on the tip. This control reduces the chance of the solder spilling over to nearby components and causing short circuits. Maintaining the proper tinning on the iron or gun, however, may be made easier by tinning with silver solder (a composition of silver, copper, and zinc). The temperature at which the bond is formed between the copper tip and the silver solder is much higher than with lead-tin solder. This tends to decrease the pitting action of the solder on the copper tip.

Overheating small or delicate wiring can easily occur when a soldering gun is used. For most jobs, even the LOW position of the trigger overheats the gun after 10 seconds. With practice, the heat can be controlled by pulsing the gun on and off with its trigger. The HIGH position is used only for fast heating and for soldering heavy connections.

When a soldering iron or gun is used, heating and cooling cycles tend to loosen the nuts or screws that hold the replaceable tips. When the nut on a gun becomes loose, the resistance of the tip connection increases. The temperature of the connection is increased, thus reducing the heat at the tip. Continued loosening may eventually cause an open circuit. Therefore, check and tighten the nut or screw, as needed.

CAUTION

Soldering guns should never be used to solder electronic components, such as resistors, capacitors, and transistors, because the heat generated can destroy the components. They should be used only on terminals, splices, and connectors (not the miniature type).

Q30. What happens if a soldering gun switch is pressed for periods longer than 30 seconds?

Q31. What causes the nuts or screws that hold the tips on soldering irons and guns to loosen?

Q32. A soldering gun should NOT be used on what components?

Resistance Soldering Set

A time-controlled resistance soldering set (figure 2-32) is now used at many maintenance activities. The set consists of a transformer that supplies 3 or 6 volts at a high current to stainless steel or carbon tips. The transformer is turned ON by a foot switch and OFF by an electronic timer. The timer can be adjusted for as long as 3 seconds soldering time. This set is especially useful for soldering cables to plugs and similar connectors; even the smallest types.

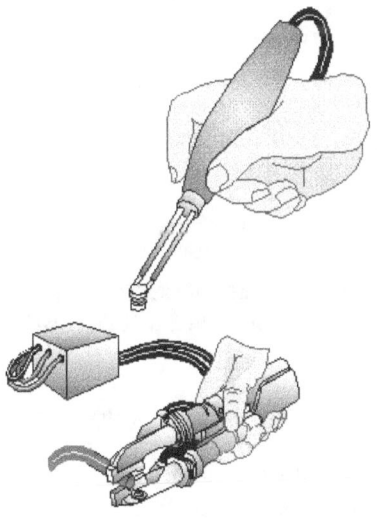

Figure 2-32.—Resistance soldering set

In use, the double-tip probes of the soldering unit are adjusted to straddle the connector cup (connector barrel) to be soldered. One pulse of current heats it for tinning. After the wire is inserted, a second pulse of current solders the connection and completes the job. Since the soldering tips are hot only during the brief period of actual soldering, burning of wire insulation and melting of connector inserts are greatly reduced.

The greatest difficulty with this device is keeping the probe tips free of rosin and corrosion. A cleaning block is mounted on the transformer case for this purpose. Some technicians prefer fine sandpaper for cleaning the double tips.

CAUTION

Do not use steel wool for cleaning tips. It is dangerous when used around electrical equipment because the strands can fall into the equipment and cause short circuits.

Q33. What is an advantage of using a resistance soldering iron when soldering wire to a connector?

Q34. Why is steel wool NEVER used as an abrasive to clean soldering tools?

Pencil Iron and Special Tips

An almost indispensable item is the pencil-type soldering iron with an assortment of tips (figure 2-33). Miniature soldering irons have a wattage rating of less than 40 watts. They are easy to use, and are recommended for soldering small components, such as miniature connectors.

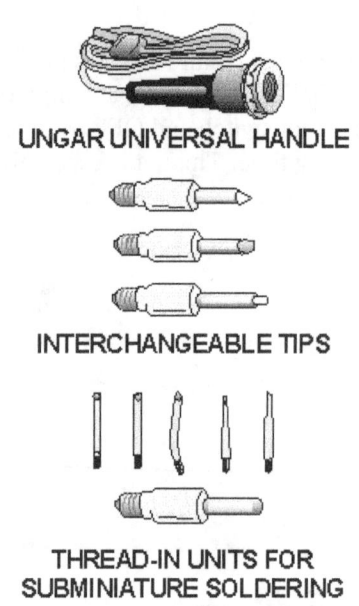

UNGAR UNIVERSAL HANDLE

INTERCHANGEABLE TIPS

THREAD-IN UNITS FOR
SUBMINIATURE SOLDERING

Figure 2-33.—Pencil iron with special tips.

One type of pencil iron is equipped with several different tips that range from one-fourth to one-half inch in size (diameter) and are of various shapes. This feature makes it adaptable to a variety of jobs. Unlike most tips that are held in place by setscrews, these tips have threads and screw into the barrel. This feature provides excellent contact with the heating element, thus improving heat transfer efficiency. "Antiseize" compound is generally applied to the threads of the tip each time a tip is installed into the iron. This allows the tip to be easily removed when another is to be inserted.

A special feature of this iron is the soldering pot that screws in like a tip and holds about a thimbleful of solder. It is useful for tinning the ends of a large number of wires.

The interchangeable tips are of various sizes and shapes for specific uses. Extra tips can be obtained and shaped to serve special purposes. The thread-in units are useful in soldering small items.

Another advantage of the pencil soldering iron is that it can be used as an improvised light source to inspect the completed work. Simply remove the soldering tip and insert a 120-volt, 6-watt, type 6S6 lamp bulb into the socket.

If leads, tabs, or small wires are bent against a board or terminal, slotted tips are provided to simultaneously melt the solder and straighten the leads.

If no suitable tip is available for a particular operation, an improvised tip can be made (see figure 2-34). Wrap a length of bare copper wire around one of the regular tips and bend the wire into the proper shape for the purpose. This method also serves to reduce thermal inertia when a larger iron must be used on small components.

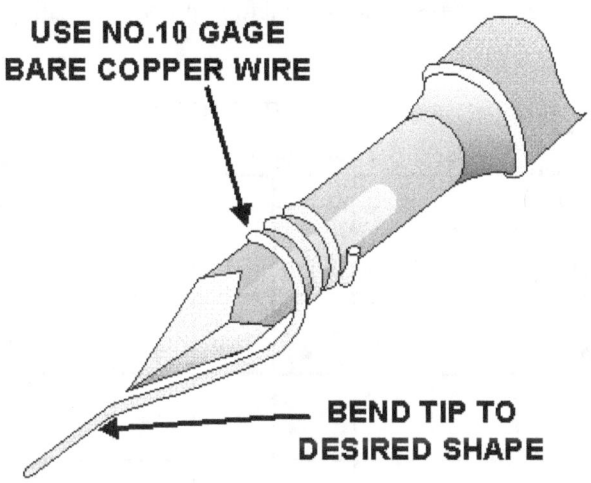

USE NO.10 GAGE
BARE COPPER WIRE

BEND TIP TO
DESIRED SHAPE

Figure 2-34.—Improvised tip.

Q35.   Why should "antiseize" compound be used on the screw-in tips of the pencil iron?

Q36.   If no suitable tip is available for a particular job, how may one be improvised?

SOLDER

Any discussion of soldering techniques should include an explanation of solder itself. Ordinary soft solder is a fusible alloy consisting chiefly of tin and lead. It is used to join two or more metals at temperatures below their melting point. In addition to tin and lead, soft solders occasionally contain varying amounts of antimony, bismuth, cadmium, or silver. These are added to change the melting point or physical properties of the alloy. Ordinary table salt has to be heated to 1,488° F before it melts. However, when a little water is added, it dissolves easily at room temperature. The action of molten solder on a metal like copper may be compared to the action of water on salt.

The solder bonds the connection by dissolving a small amount of the copper at temperatures quite below its melting point. Thus, the soldering process involves a metal solvent action between the solder and the metal being joined. A solder joint is therefore chemical in nature rather than purely physical. The bond is formed in part by chemical action and part by a physical bond.

The properties of a solder joint are different from those of the original solder. The solder is converted to a new and different alloy through the solvent action. Two metals soldered together behave like one solid metal. It is unlike two metals bolted, wired, or otherwise physically attached. These types of connections are still two pieces of metal. They are not even in direct contact due to an insulating film of oxide on the surfaces of the metals.

Temperature change does not affect the solder alloy. It withstands stress and strains without damaging the joint. An unsoldered connection eventually becomes loosened by small movements caused by temperature variations and by the gradual buildup of oxides on the metal surfaces.

To understand fully the alloy or solvent action on molten solder, look at the tin-lead fusion diagram shown in figure 2-35. This diagram shows that pure lead (point A) melts at 621° F. Point C shows the lowest melting point of the tin and lead alloy. The alloy at point C consists of 63-percent tin (SN63) and 37-percent lead. This is commonly called 63/37 solder. It has a melting point of 361° F. This type of solder, because of its very low melting point, is used in printed circuit boards and microminiature electronic repair. As you can see from the chart, the melting point of the alloy is lowered when tin is added to lead.

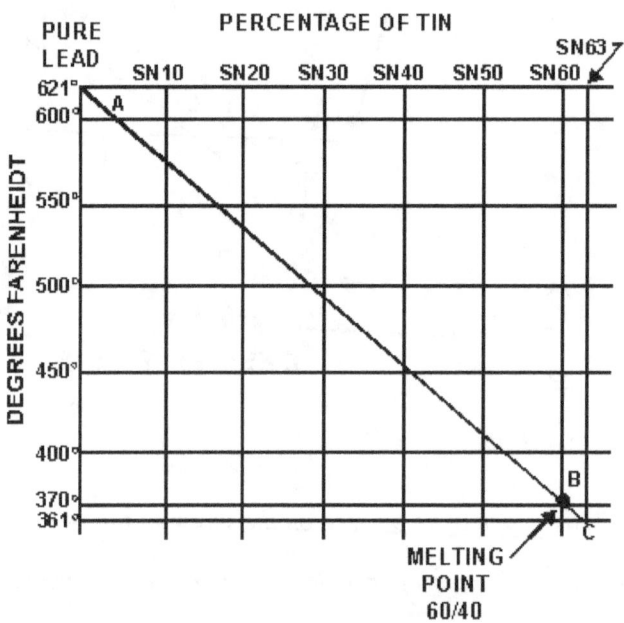

Figure 2-35.—Tin-lead fusion diagram.

The solder used to solder wires to electrical connectors, splices, and terminal lugs is a combination of 60-percent tin to 40-percent lead (60/40 solder). The melting point of 60/40 solder is 370° F, as shown at point B of the figure. Type 60/40 solder is less expensive than 63/37 solder and is suitable for all general uses.

Q37. What two metals are used to from soft solder?

Q38. Define the metal solvent action that takes place when copper conductors are soldered together.

Q39.   What is the tin-lead alloy percentage of solder used for electrical connectors, splices, and terminal lugs?

FLUX

As you know, flux is a cleaning agent to remove oxidation during soldering. Heating a metal causes rapid oxidation. Oxidation prevents solder from reacting chemically with a metal. Flux cleans the metal by removing the oxide layer. This operation is shown in figure 2-36. As the iron is moved in the direction shown, the boiling flux floats away the oxide film. The molten solder following the iron then fuses rapidly with the clean surface of the metal.

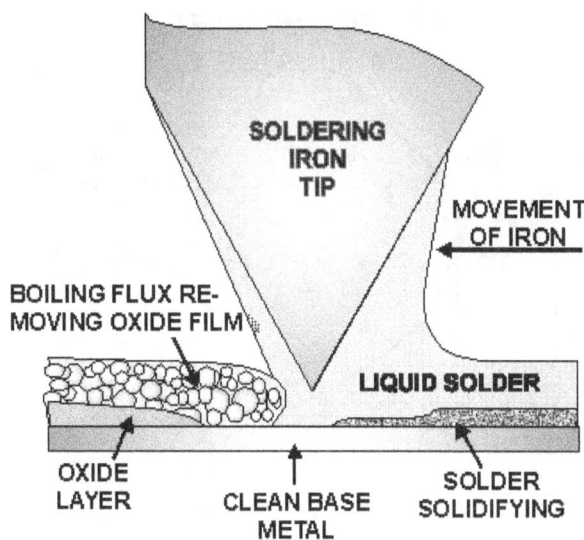

Figure 2-36.—Action of flux.

There are two classes of flux: corrosive and noncorrosive. Zinc chloride, hydrochloric acid, and sal ammoniac are corrosive fluxes. Corrosive flux should NEVER be used in electrical or electronic repair work. Use only rosin fluxes. Any flux remaining in the joint corrodes the connection and creates a defective circuit. Rosin is a noncorrosive flux and is available in paste, liquid, or powder form.

SOLVENTS

A solvent is used for cleaning and removing contaminants (oil, grease, dirt, and so forth) from the soldered connection. Solvents must be nonconductive and noncorrosive. Solvents must be used in a manner that keeps dissolved flux residue from "contact" surfaces, such as those in switches, potentiometers, or connectors. Ethyl and isopropyl alcohol are acceptable solvents.

WARNING

These cleaning solvents are highly flammable and may give off toxic vapors. Follow Navy safety precautions and take extreme care when using any flammable solvent.

Q40.   What purpose does flux serve in the soldering process?

Q41.   What type of flux must be used in all electrical and electronic soldering?

Q42.   Why are solvents used in the soldering process?

Soldering Aids

Some type of heat shunt must be used in all soldering operations that involve heat-sensitive components. A typical heat shunt (figure 2-37) permits soldering the leads of component parts without overheating the part itself. The heat shunt should be attached carefully to prevent damage to the leads, terminals, or component parts. The shunt should be clipped to the lead, between the joint and the part being protected. As the joint is heated, the shunt absorbs the excess heat before it can reach the part and cause damage.

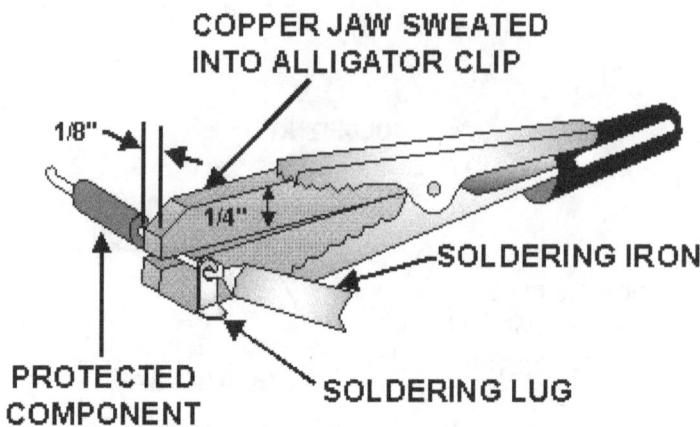

Figure 2-37.—Heat shunt.

A small piece of beeswax may be placed between the protected unit and the heat shunt. When the beeswax begins to melt, the temperature limit has been reached. The heat source should be removed immediately, but the shunt should be left in place.

Removing the shunt too soon permits the heat to flow from the melted solder into the component. The shunt should be allowed to remain in place until it cools to room temperature. A clip-on shunt is preferred because it requires positive action for removal. It does not require that the technician maintain pressure to hold it in place. This leaves both hands free to solder the connection.

Two safety devices are shown in figure 2-38. These devices prevent burns to the operator when the soldering iron is not in use for short periods of time.

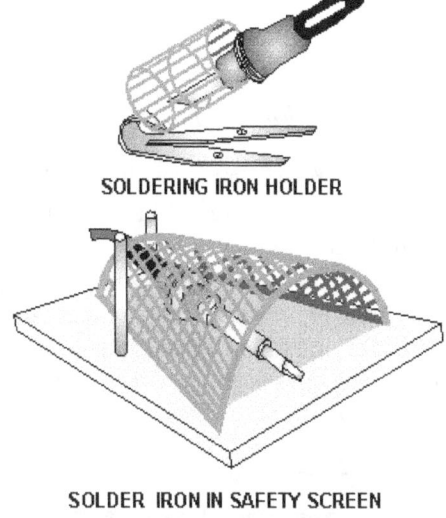

SOLDERING IRON HOLDER

SOLDER IRON IN SAFETY SCREEN

Figure 2-38.—Soldering iron safety devices.

Q43.   What is the purpose of a heat shunt?

LACING CONDUCTORS

Conductors within equipment must be kept in place to present a neat appearance and aid in tracing the conductors when alterations or repairs are required. This is done by LACING the conductors into wire bundles called cables. An example of lacing is shown in figure 2-39. When conductors are properly laced, they support each other and form a neat, single cable.

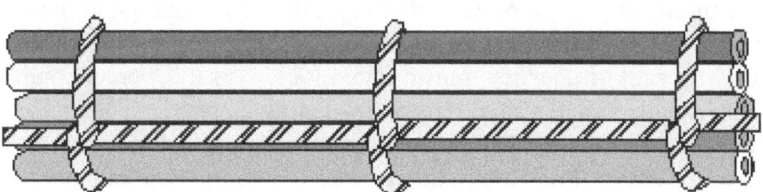

Figure 2-39.—Conductor lacing.

A narrow, flat tape should be used wherever possible for lacing and tying. This tape is not an adhesive type of tape. Round cord may also be used, but its use is not preferred because cord has a tendency to cut into wire insulation. Use cotton, linen, nylon, or glass fiber cord or tape, according to the temperature requirements. Cotton or linen cord or tape must be prewaxed to make it moisture and fungus resistant. Nylon cord or tape may be waxed or unwaxed; glass fiber cord or tape is usually not waxed.

The amount of flat tape or cord required to single lace a group of conductors is about two and one-half times the length of the longest conductor in the group. Twice this amount is required if the conductors are to be double laced.

Before lacing, lay the conductors out straight and parallel to each other. Do not twist them together because twisting makes conductor lacing and wire tracing difficult during troubleshooting.

2-37

Q44. Besides presenting a neat appearance and supporting each other, what is the other purpose for lacing conductors?

Q45. Why is flat tape preferred instead of round cord when wire bundles are laced?

Q46. What amount of flat tape or round cord is required to single lace a group of conductors?

A lacing shuttle on which the cord can be wound keeps the cord from fouling during the lacing operation. A shuttle similar to the one shown in figure 2-40 can easily be made from aluminum, brass, fiber, or plastic scrap. Rough edges of the material used for the shuttle should be filed smooth to prevent injury to the operator and damage to the cord. To fill the shuttle for a single lace, measure the cord, cut it, and wind it on the shuttle. For double lace, proceed as before, except double the length of the cord before you wind it on the shuttle. For double lace, start both ends of the cord or tape on the shuttle in order to leave a loop for starting the lace. This procedure is explained later in the chapter.

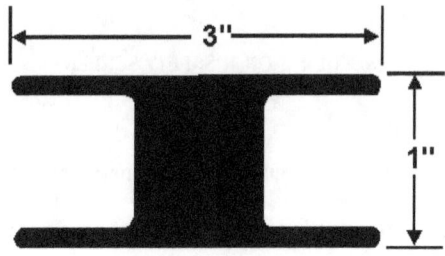

Figure 2-40.—Lacing shuttle.

Some equipment requires the use of twisted wires. One example is the use of "twisted pairs" for the ac filament leads of certain electron tube amplifiers to minimize radiation of their magnetic field. This prevents an annoying hum in the amplifier output. You should duplicate the original layout when relacing any wiring harness.

Lace or tie bundles tightly enough to prevent slipping, but not so tightly that the cord or tape cuts into or deforms the insulation. Be especially careful when lacing or tying coaxial cable. Coaxial cable is a conductor used primarily for radio-frequency transmission. It consists of a center conductor separated from an outer conductor (usually called a shield) by an insulating dielectric. The dielectric maintains a constant capacitance between the two conductors, which is very important in radio transmission. The dielectric is soft and deforms easily if tied too tightly or with the wrong type of tape.

CAUTION

Do not use round cord for lacing or tying coaxial cable or bundles that contain coaxial cable. Use only the approved military specification tape to lace or tie coaxial cables or bundles containing coaxial cables.

Q47. What is the purpose of a lacing shuttle?

Q48. When should wires be twisted prior to lacing?

Q49. What precautions should you take when tying bundles containing coaxial cables?

SINGLE LACE

Single lace can be started with a square knot and at least two marling hitches drawn tightly. Details of the square knot and marling hitch are shown in figure 2-41. Do not confuse the marling hitch with a half hitch. In the marling hitch, the end is passed over and under the strand, as shown in view A of the figure. After forming the marling hitches, draw them tightly against the square knot, as shown in view B. The lace consists of a series of marling hitches evenly spaced at 1/2-inch to 1-inch intervals along the length of the group of conductors, as shown in view C of the figure.

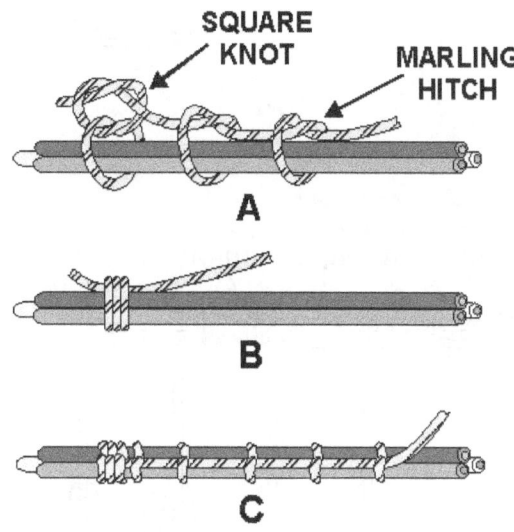

Figure 2-41.—Applying single lace.

When dividing conductors to form two or more branches, follow the procedure illustrated in figure 2-42. Bind the conductors with at least six turns between two marling hitches, and continue the lacing along one of the branches, as shown in view A. Start a new lacing along the other branch. To keep the bends in place, form them in the conductors before lacing. Always add an extra marling hitch just prior to a breakout as shown in view B.

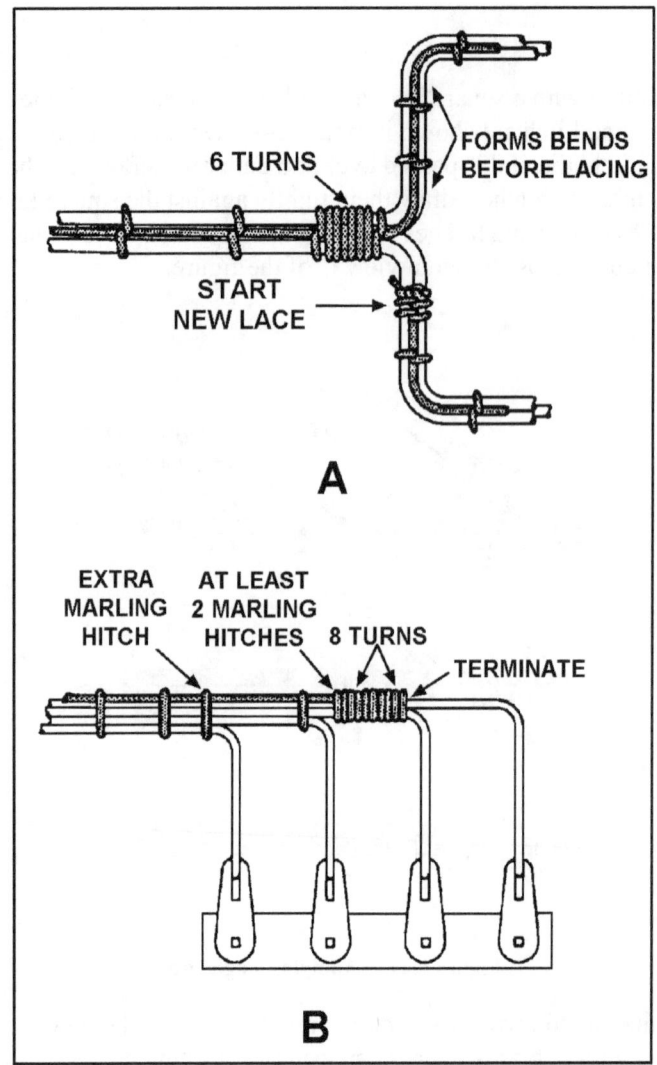

Figure 2-42.—Lacing branches and breakouts.

Double lace should be used on groups of conductors that are 1 inch or larger in total diameter. Either a single lace or a double lace may be used on groups of less than 1 inch.

Q50.   How is the single lace started?

DOUBLE LACE

Double lace is applied in a manner similar to single lace, except that it is started with a telephone hitch and is double throughout the length of the lacing (figure 2-43). Both double and single lace may be ended by forming a loop from a separate length of cord and using it to pull the end of the lacing back underneath a serving of approximately eight turns (figure 2-44). An alternate method of ending the lacing is illustrated in figure 2-45. This method can also be used for either single- or double-cord lacing. Another method is by using a marling hitch as a lock stitch (figure 2-46) to prevent slippage. This procedure will also prevent unraveling should a break occur to the lacing.

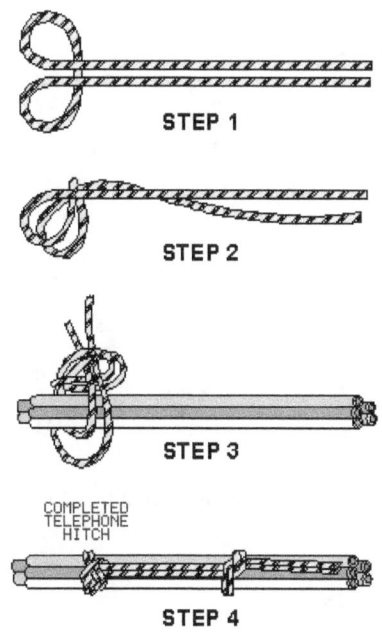

STEP 1

STEP 2

STEP 3

COMPLETED
TELEPHONE
HITCH

STEP 4

Figure 2-43.—Starting double lace.

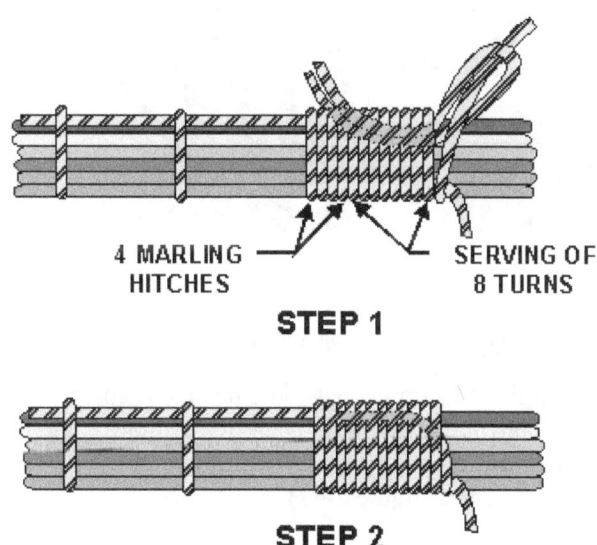

4 MARLING
HITCHES

SERVING OF
8 TURNS

STEP 1

STEP 2

Figure 2-44.—Terminating double lace.

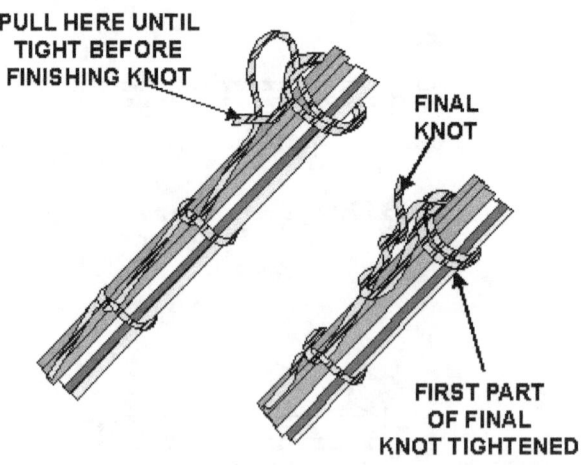

PULL HERE UNTIL
TIGHT BEFORE
FINISHING KNOT

FINAL
KNOT

FIRST PART
OF FINAL
KNOT TIGHTENED

Figure 2-45.—Alternate method of terminating the lace.

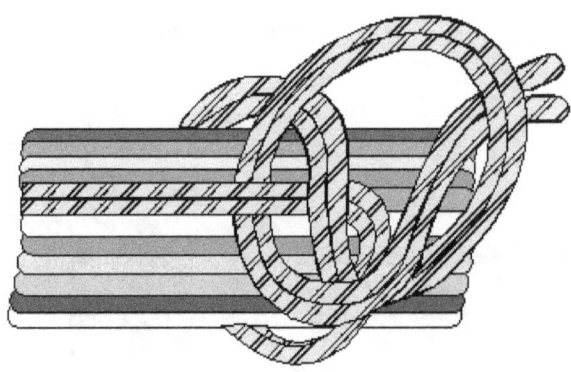

Figure 2-46.—Marling hitch as a lock stitch.

The spare conductors of a multiconductor cable should be laced separately, and then tied to active conductors of the cable with a few telephone hitches. When two or more cables enter an enclosure, each cable group should be laced separately. When groups are parallel to each other, they should be bound together at intervals with telephone hitches (figure 2-47).

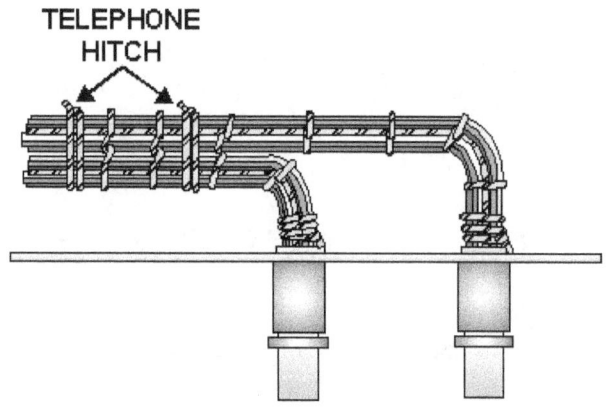

Figure 2-47.—Spot tying cable groups.

Q51. What size wire bundles require double lace?

Q52. How is the double lace started?

Q53. How are laced cable groups bound together?

SPOT TYING

When cable supports are used in equipment as shown in figure 2-48, spot ties are used to secure the conductor groups if the supports are more than 12 inches apart. The spot ties are made by wrapping the cord around the group as shown in figure 2-49. To finish the tie, use a clove hitch followed by a square knot with an extra loop. The free ends of the cord are then trimmed to a minimum of 3/8 inch.

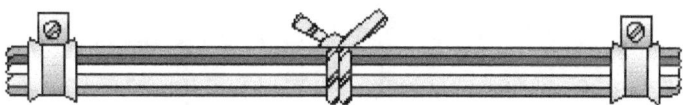

Figure 2-48.—Use of spot ties.

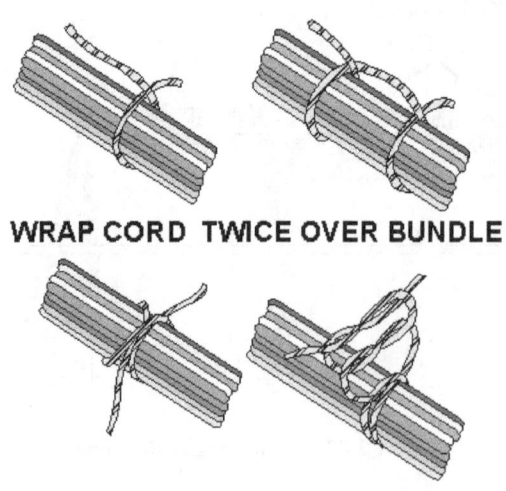

WRAP CORD TWICE OVER BUNDLE

CLOVE HITCH & SQUARE KNOT

Figure 2-49.—Making spot ties.

SELF-CLINCHING CABLE STRAPS

Self-clinching cable straps are adjustable, lightweight, flat nylon straps. They have molded ribs or serrations on the inside surface to grip the wire. They may be used instead of individual cord ties for securing wire groups or bundles quickly. The straps are of two types: a plain cable strap and one that has a flat surface for identifying the cables.

CAUTION

Do not use nylon cable straps over wire bundles containing coaxial cable. Do not use straps in areas where failure of the strap would allow the strap to fall into movable parts.

Installing self-clinching cable straps is done with a Military Standard hand tool, as shown in figure 2-50. An illustration of the working parts of the tool is shown in figure 2-51. To use the tool, follow the manufacturer's instructions.

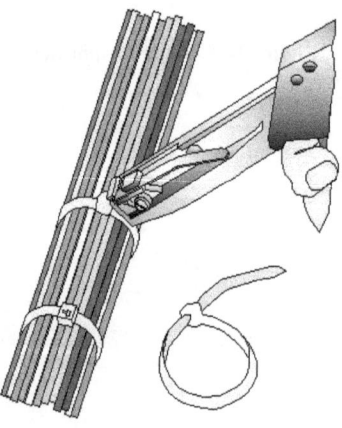

Figure 2-50.—Installing self-clinching cable straps.

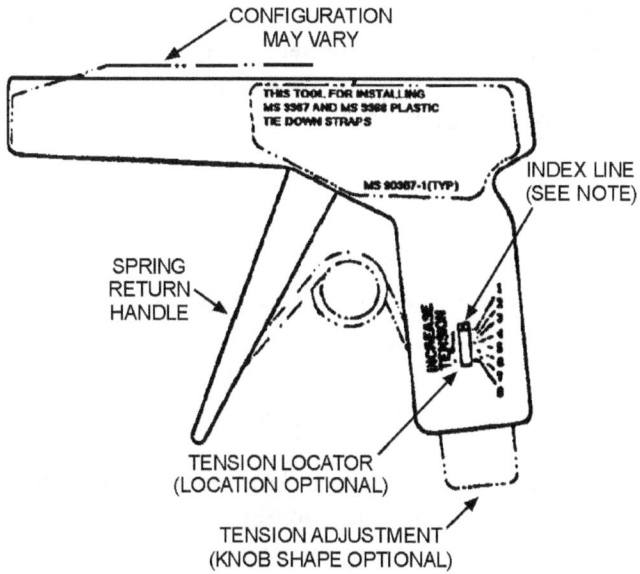

Figure 2-51.—Military Standard hand tool for self-clinching cable straps.

WARNING

Use proper tools and make sure the strap is cut flush with the eye of the strap. This prevents painful cuts and scratches caused by protruding strap ends. Do not use plastic cable straps in high-temperature areas (above 250° F).

HIGH-TEMPERATURE PRESSURE-SENSITIVE TAPE LACING

High-temperature, pressure-sensitive tape must be used to tie wire bundles in areas where the temperature may exceed 250° F. Install the tape as follows (figure 2-52):

1. Wrap the tape around the wire bundle three times, with a two-thirds overlap for each turn.

2. Heat-seal the loose tape end with the side of a soldering iron tip.

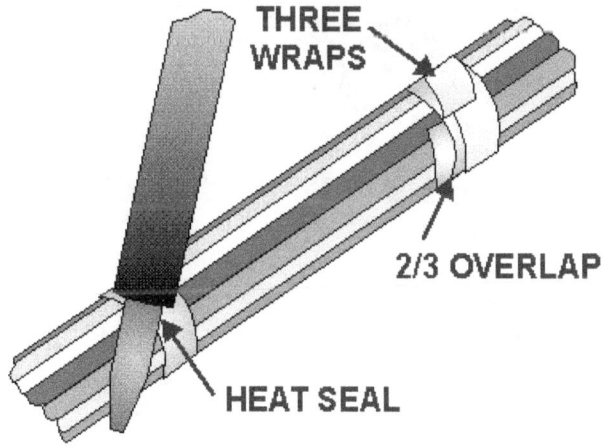

Figure 2-52.—Securing wire bundles in high-temperature areas.

WARNING

Insulation tape (including the glass fiber type) is highly flammable and should not be used in a high-temperature environment. Only insulation tape approved for high-temperature operation (suitable for continuous operation at 500° F) should be used in high-temperature environments.

Q54.   When are spot ties used?

Q55.   What is used to install self-clinching cable straps?

Q56.   What is used to tie wire bundles in high-temperature areas?

SUMMARY

In this chapter you have learned some of the basic skills required for proper wiring techniques. We have discussed conductor splices and terminal connections, basic soldering skills, and lacing and tying wire bundles.

The basic requirement for any splice or terminal connection is that it be both mechanically and electrically as strong as the conductor or device with which it is to be used.

Insulation Removal—The first step in splicing or terminating electrical conductors is to remove the insulation. The preferred method for stripping wire is by use of a wire-stripping tool. The hot-blade stripper cannot be used on such insulation material as glass braid or asbestos. An alternate method for stripping copper wire is with a knife. A knife is the required tool to strip aluminum wire. Take extreme care when stripping aluminum wire. Knicking the strands will cause them to break easily.

Western Union Splice—A simple connection known as the Western Union splice is used to splice small, solid conductors together. After the splice is made, the ends of the wire are clamped down to prevent damage to the tape insulation.

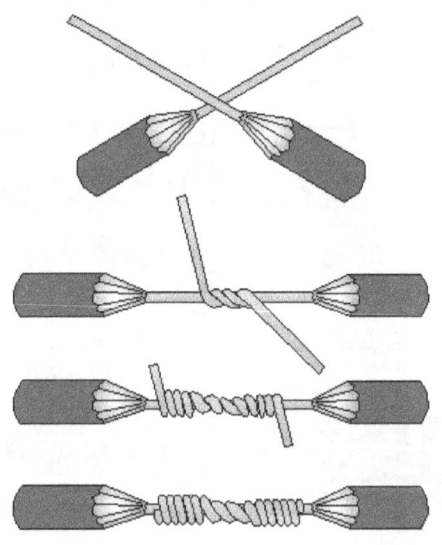

Staggered Splice—The staggered splice is used on multiconductor cables to prevent the joint from being bulky.

Rattail Joint—A splice that is used in a junction box and for connecting branch circuits; wiring is placed inside conduits.

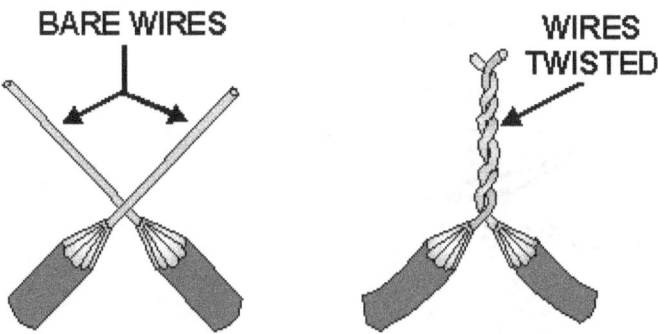

Fixture Joint—When conductors of different sizes are to be spliced, such as fixture wires to a branch circuit, the fixture joint is used.

Knotted Tap Joint—This type of splice is used to splice a conductor to a continuous wire. It is not considered a "butted" splice as the ones previously discussed.

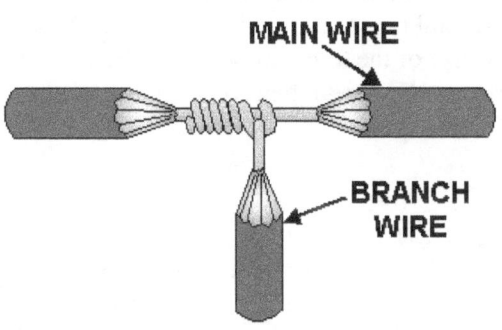

Splice Insulation—Rubber tape is an insulator for the type of splices we have discussed so far.

Friction Tape—It has very little insulating value but is used as a protective covering for the rubber tape. Another type of insulating tape is plastic electrical tape, which is quite expensive.

Terminal Lugs—The terminals used in electrical wiring are either of the soldered or crimped type. The advantage of using a crimped type of connection is that it requires very little operator skill, whereas the soldered connection is almost completely dependent on the skill of the operator. Some form of insulation must be used with noninsulated splices and terminal lugs. The types used are clear plastic tubing (spaghetti) and heat-shrinkable tubing. When a heat gun is used to shrink the heat-shrinkable tubing, the maximum allowable heat to be used is 300° F. When using the compressed air/nitrogen heating tool, the air/nitrogen source cannot be greater than 200 psig.

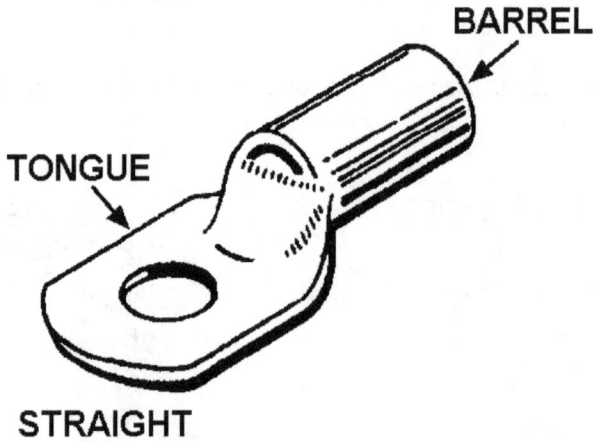

BARREL

TONGUE

STRAIGHT

Aluminum Terminals and Splices—Aluminum terminals and splices are noninsulated and very difficult to use. Some of the things you should remember when working with aluminum wire are: (1) Never attempt to clean the aluminum wire. There is a petroleum abrasive compound in the terminal lug or splice that automatically cleans the wire. (2) The only tools that should be used for the crimping operation are the power crimping type. (3) Never use lock washers next to aluminum terminal lugs as they will gouge out the tinned area and increase deterioration.

Preinsulated Copper Terminal Lugs and Splices—The most common method of terminating and splicing copper wires is with the use of preinsulated terminal lugs and splices. Besides not having to insulate the terminal or splice after the crimping operation, the other advantage of this type is that it gives extra wire insulation support. Several types of crimping tools can be used for these types of terminals and splices. The tool varies with the size of the terminal or splice. Preinsulated terminal lugs and splices are color coded to indicate the wire size they are to be used with.

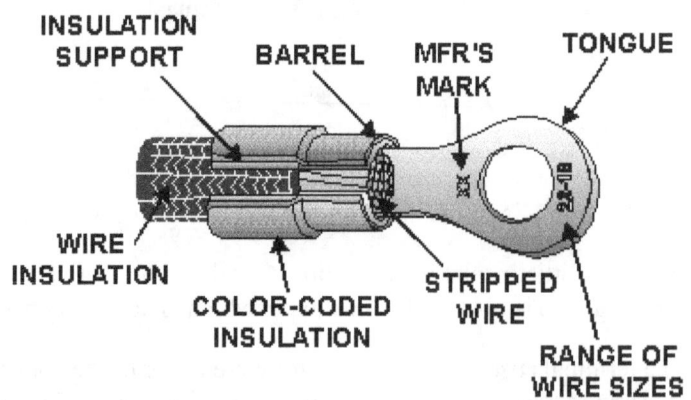

INSULATION SUPPORT   BARREL   MFR'S MARK   TONGUE

WIRE INSULATION   COLOR-CODED INSULATION   STRIPPED WIRE   RANGE OF WIRE SIZES

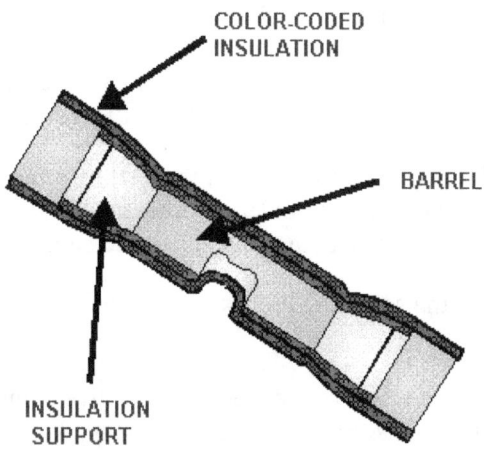

COLOR-CODED
INSULATION

BARREL

INSULATION
SUPPORT

Soldering—The basic skills required to solder terminal lugs, splices, and electrical connectors are covered in this area. Prior to any soldering operation, the items to be soldered must be cleaned; they will not adhere to dirty, greasy, or oxidized surfaces. The next step is the "tinning" process. This process is accomplished by coating the material to be soldered with a bright coat of solder. The wire to be soldered must be stripped to 1/32 inch longer than the depth of the solder cup of the terminal, splice, or connector to which it is to be soldered. This is to prevent burning the insulation. It also allows the wire to flex at the stress point. When you tin the wire, it should be done to one-half of the stripped length. When soldering a connection, take precaution to prevent movement of the parts while the solder is cooling. A "fractured solder" joint will result if this precaution is not taken.

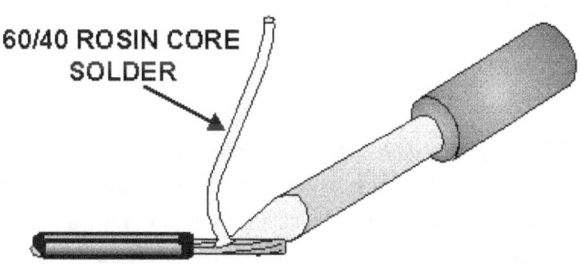

60/40 ROSIN CORE
SOLDER

Soldering Tools—The important difference in soldering iron sizes is not the temperature (they all produce 500° F to 600° F), but the thermal inertia. Thermal inertia is the ability of soldering tools to maintain a satisfactory soldering temperature while giving up heat to the joint to be soldered. A well-designed soldering iron is self-regulating because its heating element increases with the rising temperature, thus inciting the current to a satisfactory level. When using a soldering gun, do not press the switch for periods longer than 30 seconds. Doing so will cause the tip to overheat to the point of incandescence. The nuts or screws that retain the tips on soldering irons and guns tend to loosen because of the continuous heating and cooling cycles. Therefore, they should be tightened periodically. You should <u>never</u> use a soldering gun on electronics components, such as resistors, capacitors, or transistors. An advantage of using a resistance soldering iron to solder a wire to a connector is that the soldering tips are only hot during the brief period of soldering the connection.

Solder—Ordinary soft solder is a fusible alloy of tin and lead used to join two or more metals at temperatures below their melting point. The metal solvent action that occurs when copper conductors are soldered together takes place because a small amount of the copper combines with the solder to form a new alloy. Therefore, the joint is one common metal. The tin-lead alloy used for general-purpose soldering is composed of 60-percent tin and 40-percent lead (60/40 solder).

Flux—Flux is used in the soldering process to clean the metal by removing the oxide layer on the metal and to prevent further oxidation during the soldering process. Always use noncorrosive, nonconducting rosin fluxes when soldering electrical and electronic components.

Solvents—Solvents are used in the soldering process to remove contaminants from the surfaces to be soldered.

Soldering Aids—Use a heat shunt when you solder heat-sensitive components. It dissipates the heat, thereby preventing damage to the heat-sensitive component. Some type of soldering iron holder or guard should be used to prevent the operator from being burned.

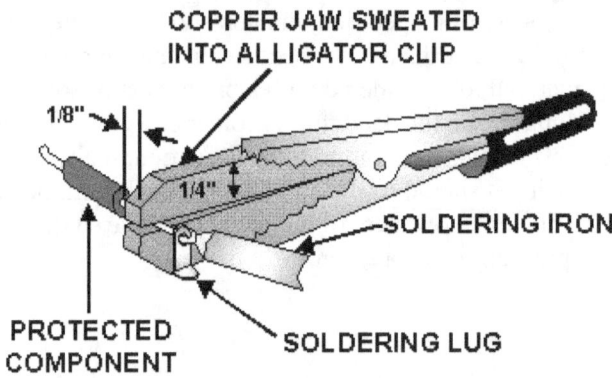

Lacing Conductors—The purpose of lacing conductors is to present a neat appearance and to facilitate tracing the conductors when alterations or repairs are required. Flat tape is preferred for lacing instead of round cord. Cord has a tendency to cut into the wire insulation. The amount of flat tape or round cord required to lace a group of conductors is about two and one-half times the length of the longest conductor. A lacing shuttle is useful during the lacing operation to prevent the tape or cord from fouling. Wires should only be twisted prior to lacing if it is required, such as for filament leads in electron tube amplifiers. When lacing wire bundles containing coaxial cables, use the proper flat tape and do not tie the bundles too tightly. Never use round cord on coaxial cable. A single lace is started with a square knot and at least two marling Hitches. A double lace is required for wire bundles that are 1 inch or more in diameter. It is started with a telephone hitch. Cable groups are bound together by use of telephone hitch

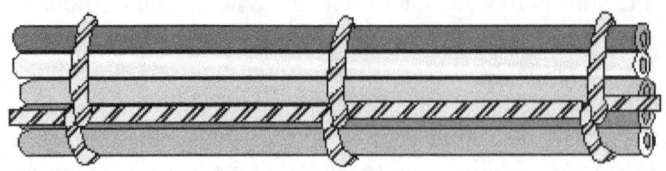

Spot Ties—Spot ties are used when cable supports are used that are more than 12 inches apart.

Self-clinching Cable Straps—If self-clinching cable straps are used, they should be installed with the Military Standard hand tool designed for their use.

High-temperature Areas—When you are required to tie wire bundles in high-temperature operating areas, use only high-temperature, pressure-sensitive tape.

ANSWERS TO QUESTIONS Q1 THROUGH Q56.

A1.   The connection must be both mechanically and electrically as strong as the conductor or device with which it is used

A2.   By use of a wire-stripping tool

A3.   Hot-blade stripper.

A4.   Knife.

A5.   To prevent damage to the tape insulation.

A6.   To prevent the joint from being bulky.

A7.   When wires are in conduit and a junction box is used.

A8.   Fixture joint.

A9.   Knotted tap joint.

A10. As a protective covering over the rubber tape.

A11. Requires relatively little operator skill to install.

A12. Spaghetti or heat-shrinkable tubing.

A13. 300° F

A14. 200 psig.

A15. No, it is done automatically by the petroleum abrasive compound that comes in the terminal or splices.

A16. Power-operated crimping tools.

A17. It gouges the terminal lug and causes deterioration.

A18. The use of preinsulated splices and terminal lugs.

A19. It has insulation support for extra supporting strength of the wire insulation.

A20. To identify wire sizes they are to be used on.

A21. Solder will not adhere to dirty, greasy, or oxidized surfaces.

A22. The coating of the material to be soldered with a light coat of solder.

A23. To prevent burning the insulation during the soldering process and to allow the wire to flex easier at a stress point.

A24. One-half the stripped length.

A25. Movement of the parts being soldered while the solder is cooling.

A26. The capacity of the soldering iron to generate and maintain a satisfactory soldering temperature while giving up heat to the joint being soldered.

A27. Although its temperature is as high as the larger irons, it does not have thermal inertia.

A28. The resistance of its heating element increases with rising temperature, thus limiting the current flow.

A29. File the tip until it is smooth and retin it.

A30. It will overheat and could burn the insulation of the wire being soldered.

A31. The heating and cooling cycles.

A32. Electronic components, such as resistors, capacitors, and transistors.

A33. The soldering tips are hot only during the brief period of soldering the connection, thus minimizing the chance of burning the wire insulation or connector inserts.

A34. The strands can fall into electrical equipment being worked on and cause short circuits.

A35. It enables the tip to be removed easily when another is to be inserted.

A36. Wrap a length of copper wire around one of the regular tips and bend to the proper shape for the purpose.

A37. Tin and lead.

A38. The solder dissolves a small amount of the copper, which combines with the solder forming a new alloy; therefore, the joint is one common metal.

A39. 60-percent tin and 40-percent lead (60/40 solder).

A40. It cleans the metal by removing the oxide layer and prevents further oxidation during the soldering.

A41. Noncorrosive, nonconductive rosin fluxes.

A42. To remove contaminants from soldered connections.

A43. To prevent damage to heat-sensitive components.

A44. To aid in tracing the conductors when alterations or repairs are required.

A45. Round cord has a tendency to cut into the wire insulation.

A46. Two and one-half times the length of the longest conductor in the group.

A47. To keep the tape or cord from fouling during the lacing operation.

A48. When required, such as for the filament leads in electron tube amplifiers.

A49. Do not tie too tightly and use the proper type of tape.

A50. With a square knot and at least two marling hitches drawn tightly.

A51. Bundles that are 1 inch or larger in diameter

A52. With a telephone hitch.

A53. They are bound together at intervals with telephone hitches.

A54. When wire bundles are supported by cable supports that are more than 12 inches apart.

A55. Military Standard hand tool.

A56. High-temperature, pressure-sensitive tape.